ENERGETIC PRINCIPLES OF CHEMICAL REACTIONS

ENERGETIC PRINCIPLES
OF CHEMICAL REACTIONS

Jack Simons
UNIVERSITY OF UTAH

JONES AND BARTLETT PUBLISHERS, INC.
BOSTON   PORTOLA VALLEY

Copyright © 1983 by Jones and Bartlett Publishers, Inc. All rights reserved. No part of the material protected by this copyright notice may be reproduced or utilized in any form, electronic or mechanical, including photocopying, recording, or by any information storage and retrieval system, without written permission from the copyright owner.

Editorial offices: Jones and Bartlett Publishers, Inc., 30 Granada Court, Portola Valley, CA 94025.

Sales and customer service offices: Jones and Bartlett Publishers, Inc., 20 Park Plaza, Boston, MA 02116.

Library of Congress Cataloging in Publication Data

Simons, Jack.
   Energetic principles of chemical reactions.
   Includes bibliographical references and index.
   1. Chemical reaction, Conditions and laws of.
I. Title.
QD501.S584   1983   541.3'9   82-23432
ISBN 0-86720-020-0

ISBN: 0-86720-020-0

Publisher: Arthur C. Bartlett
Book and Cover Design: Hal Lockwood
Manuscript Editor: David Freifelder
Production: Robin Lockwood
Composition: Mariner Typographers
Printer and Binder: The Alpine Press

Printed in the United States of America
Printing number (last digit) 10 9 8 7 6 5 4 3 2 1

TO MY MOTHER,
Mrs. M. A. Simons

Contents

Part 1 Underlying Physical Principles   1

    Chapter 1   Potential Energy Surfaces   3
        1.1   Born-Oppenheimer and Adiabatic Approximations   6
        1.2   Intersections of Potential Energy Surfaces   8

    Chapter 2   Symmetry and Potential Energy Surfaces   11
        2.1   Slope of the Energy Surface: First-Order Jahn-Teller Effect   16
        2.2   Surface Curvature: Second-Order Jahn-Teller Effect   17

    Chapter 3   Review of Molecular Orbital and Configuration-Mixing Ideas   23
        3.1   Molecular Orbitals: Symmetry of the Fock Operator   24
        3.2   When Can Orbital Energies be Added?   27
        3.3   Configuration Construction and Mixing   28
        3.4   Approximate Symmetry   30

Part 2 Applications to Thermal Reactions   32

    Chapter 4   Examples for Analyzing Ground-State Thermal Processes 35
        4.1   Simple Predictions from Orbital-, Configuration-, and State-Correlation Diagrams   35
        4.2   An Example of Unimolecular Decomposition   40
        4.3   A Degenerate Case with Jahn-Teller Effects   44
        4.4   The Bond-Symmetry Rule—Another Jahn-Teller Case   46
        4.5   Breaking of Single Homonuclear and Heteronuclear Bonds   50
        4.6   The Use of Bonding-Antibonding Orbital Mixing to Predict the Reaction Coordinate   52

viii CONTENTS

        4.7 Electrocyclic Reactions by Occupied-Orbital Following 54
        4.8 Cycloaddition Reactions by Orbital Following 58
        4.9 Sigmatropic Migrations via HOMO-LUMO Overlaps 63
        4.10 A Topology-Based Method 65

        Problems 68

Part 3 Theory and Applications Pertaining to Photochemical Processes 72

Chapter 5 Introductory Remarks about Photochemical Reactions 75
    5.1 Nature of Low-Energy Excited States 75
    5.2 Energy Redistribution in the Singlet Manifold 77
    5.3 Processes Involving Triplet States 79

Chapter 6 Internal Conversion and Intersystem Crossing 83
    6.1 The States Between Which Transitions Occur 83
    6.2 Rates of Transitions 85
    6.3 Intersystem Crossing Rates 92

Chapter 7 Examples of Photochemical Reactions 95
    7.1 Dimerization of Two Ethylenes 95
    7.2 Closure of 1,3-Butadiene to Cyclobutene 99
    7.3 HOMO-LUMO (SOMO) Overlap for the Diels-Alder Reaction 102
    7.4 Excited Reactants Can Correlate Directly with Ground-State Products 104
    7.5 Benzene Photochemistry 106
    7.6 $C + H_2 \rightarrow CH_2$ 112

    Problems 120

Appendix A Overview of *ab Initio* Molecular Orbital Theory 123
    A.1 Orbitals 123
    A.2 Configuration Interaction 128
    A.3 Slater-Condon Rules 130

Appendix B The Nature of Photon-Induced Electronic Transitions 137

Appendix C Review of Point-Group Symmetry Tools 141

Answers 155

References 159

Index 163

Preface

This text is intended for a one-quarter or one-semester course at the first- or second-year graduate level. It discusses how symmetry concepts, orbital nodal patterns, and molecular topology can be used to make statements about energetics in chemical reactions. It also differs from commonly used texts that only consider how orbital-symmetry constraints allow or forbid various reactions in its more rigorous approach. Introductory chapters explain the physical origins of orbital-, configuration-, and state-correlation diagrams, Jahn-Teller instability, internal conversion, and intersystem crossing. These sections are for students who desire a rigorous understanding of the physical origins of these concepts as they relate to thermal and photochemical processes. These discussions are not unduly long, however, and they contain sufficient physical interpretations to make them valuable reading for graduate students and researchers in all areas of chemistry. Following the introduction of the physical principles, applications to explicit thermal and photochemical reaction problems show the practical uses of these tools. These examples, which are written in a tutorial style, should appeal to all students of chemistry.

Most one-semester introductory courses on quantum chemistry and two-quarter combined courses on quantum chemistry and spectroscopy should provide adequate background to understand the material in this text. Some concepts of group theory also appear; more advanced topics are taught in the text as needed. General ideas of Hartree-Fock molecular orbital theory sometimes come into use, but only to the extent that they are absolutely necessary. Little is said about numerical application of molecular orbital methods; instead, emphasis is placed on the conceptual use of orbitals and their symmetries in chemical reactions. Three appendixes are provided for the reader to review or to learn the requisite background material dealing with *ab initio* molecular theory, molecular-point-group symmetry methods, and the photon absorption process that prepares molecular reactants for subsequent photoreaction. For the reader who wishes to test his or her mastery of the material, two sets of problems are provided.

This text is the result of a one-quarter graduate course taught to first-year graduate students in physical, organic, and inorganic chemistry at the University of Utah. I wish to thank the students who have contributed to its development. I also wish to acknowledge much helpful input provided by several of our graduate students and postdoctoral fellows—David Chuljian, Judy Ozment, Ron Shepard, and Ajit Banerjee—as well as the support and advice given by my colleagues Poul Jørgensen, Bill Breckenridge, and Josef Michl.

June 1983   Jack Simons

ENERGETIC PRINCIPLES OF CHEMICAL REACTIONS

Part 1
Underlying Physical Principles

In the first three chapters of this text we show how to express, in quantitative terms, certain concepts that are widely used in a qualitative manner in chemical education and research. Specifically, we analyze how one defines potential energy surfaces and reaction coordinates, and we examine the conditions under which these concepts break down. By observing how the reaction coordinate varies as the reaction proceeds from reactants, through one or more transition states, to products, the idea of symmetry conservation is developed. In Chapter 3 we discuss the concepts of orbitals, electronic occupancies (configurations), and electronic states, and we show why symmetry conservation applies at each of these three levels.

The ultimate goal of this text is to permit the reader to predict whether any postulated chemical reaction should experience a large activation energy barrier and, thereby, be forbidden. To make such predictions, one must be able to visualize the reactant molecules moving on a potential energy surface that is characteristic of either the ground state or an excited electronic state. Such qualitative visualization can be carried out only after one has achieved a good appreciation of the electronic structures (i.e., orbital shapes and energies, and orbital occupancies) of the reactants, products, and likely transition states. The first three chapters of this text develop these important tools.

The level of presentation in these first three chapters is substantially more sophisticated than in most other books that deal with symmetry in chemical reactions (for example, Pearson, 1976; Woodward and Hoffmann, 1970; Borden, 1975; or Fleming, 1976). This level is especially relevant to physical chemists whose research requires a *quantitative* interpretation of experimental data—modern research in chemical dynamics and spectroscopy often demands the use of such theoretical tools. Likewise, it is important also that researchers who wish to make qualitative use of symmetry ideas be aware of the origins and limitations of such concepts. Therefore, although the vast majority of the examples treated in later chapters make only qualitative use of the theoretical machinery covered in Chapters 1-3, it is essential that all modern researchers be well founded in the physical origins of these valuable symmetry tools. It is recommended that readers who are not familiar with the foundations of molecular orbital theory and point-group symmetry read Appendixes A and C before attempting to master these first three chapters.

Chapter 1

Potential Energy Surfaces

As will become clear shortly, a potential energy surface is merely a construct of one's imagination. It is an idea that has proved to be of immense value for conceptualizing chemical reactions but that loses its rigorous content in certain circumstances. Within its range of approximate validity, a potential energy surface can be thought of as the topographical map describing the terrain on which the reactant molecules must move on their route to a transition state and then onward toward the geometrical arrangement of the product molecules. To understand better what these potential energy surfaces are, it is useful to examine how they arise in the quantum mechanical treatment of the motions of the nuclei and the electrons that comprise the reactant molecules.

The Hamiltonian function describing motion of a collection of nuclei of masses M_a and charges $Z_a e$ and electrons of mass m and charge $-e$ is

$$H = \sum_a \left[-\frac{\hbar^2}{2M_a} \nabla_a^2 + \frac{1}{2} \sum_{b \neq a} \frac{Z_a Z_b e^2}{|\mathbf{R}_a - \mathbf{R}_b|} \right]$$
$$+ \sum_i \left[-\frac{\hbar^2}{2m} \nabla_i^2 + \frac{1}{2} \sum_{j \neq i} \frac{e^2}{|\mathbf{r}_i - \mathbf{r}_j|} - \sum_a \frac{Z_a e^2}{|\mathbf{r}_i - \mathbf{R}_a|} \right], \quad (1.1)$$

where $(\mathbf{R}_a, \mathbf{r}_i)$ refers to a coordinate system that is fixed in space and not on the molecule.

It is convenient to rewrite the Hamiltonian in terms of molecule-fixed coordinates instead of absolute coordinates. Two such coordinate transformations might be used. First, one could introduce the *total* center of mass

$$\mathbf{R} = \frac{1}{M} \left[\sum_a M_a \mathbf{R}_a + \sum_i m \mathbf{r}_i \right],$$

in which M is the total mass of all nuclei and electrons and coordinates relative to $\mathbf{R}$. This is a good and natural choice but one that is not convenient once the idea of clamped nuclei, which we will use repeatedly, is introduced. If the nuclei are held fixed and the electrons move, the center of mass, which is the coordinate origin, could move; as a result, what was ascribed to electronic motion would include some center-of-mass motion. As will become clear shortly, our desire to think of the nuclei as fixed is important; the clamped-nuclei concept rests at the center of our ideas of potential energy surfaces.

A second transformation uses the center of mass of the nuclei to define a molecule-fixed coordinate origin

$$\mathbf{R} = \frac{1}{M} \sum_a M_a \mathbf{R}_a; \quad M = \sum_a M_a.$$

Because the electrons are so light, this position will almost be the true center of mass, and this location will remain fixed if we later clamp the nuclei. Upon expressing the positions of the nuclei and electrons as $\mathbf{R}$ plus internal or relative position vectors (for which we now use $\mathbf{R}_a$, $\mathbf{r}_i$), the above Hamiltonian can be written, for a diatomic molecule,

$$H = \sum_i \left[\left(\frac{-\hbar^2}{2m} \nabla_i^2 \right) - \sum_a \frac{Z_a e^2}{|\mathbf{r}_i - \mathbf{R}_a|} + \frac{1}{2} \sum_{j \neq i} \frac{e^2}{|\mathbf{r}_i - \mathbf{r}_j|} \right]$$
$$+ \frac{Z_a Z_b e^2}{|\mathbf{R}_a - \mathbf{R}_b|} - \frac{\hbar^2}{2\mu} \nabla_{|\mathbf{R}_a - \mathbf{R}_b|}^2 - \frac{\hbar^2}{2M} \sum_{i,j} \nabla_i \cdot \nabla_j - \frac{\hbar^2}{2M} \nabla_\mathbf{R}^2 \quad (1.2)$$

in which μ, the reduced mass of the nuclei is $M_a M_b / M$. Pack and Hirschfelder (1967) show the details of how both this transformation and the total-center-of-mass transformation mentioned above are carried out. For a more complicated molecule, only the fourth and fifth terms would differ; the fourth would be

$$\frac{1}{2} \sum_{a,b} \frac{Z_a Z_b e^2}{|\mathbf{R}_a - \mathbf{R}_b|}$$

and the fifth would be the internal kinetic energy operator, which we label h_N, describing the vibrations and rotations of the nuclei. As an example, consider a triatomic molecule ABC. For such a molecule $3N - 3 = 6$ such coordinates are needed, and these could be the vectors $\mathbf{R}_C - \mathbf{R}_B$ and $\mathbf{R}_A - \mathbf{R}_B$ or the lengths $|\mathbf{R}_C - \mathbf{R}_B|$, $|\mathbf{R}_A - \mathbf{R}_B|$, and the angle θ_{ABC} and three Euler orientation angles. The choice is up to you and should be made to simplify the treatment of the vibration/rotation problem, which Wilson, Decius, and Cross (1955) treat in elegant detail. The seventh term in equation 1.2, the motion of the nuclear center of mass, separates exactly. Hence, this motion is uncoupled from the internal electronic and vibration/rotation motion and will therefore be assumed to have been removed from further consideration (by separation of variables).

The first four terms and the sixth term in equation 1.2 are usually combined and called the electronic Hamiltonian (h_e) because they contain differential operators only for the r_i coordinates. Notice that the electronic energy will then contain the repulsion of the nuclei, so the potential energy curves will become infinitely repulsive as two nuclei approach one another. The sixth term, sometimes called the mass-polarization term, usually has small effects because it is multiplied by the inverse of the total nuclear mass M (a small factor) in contrast with the electronic kinetic energy term, which has a $h^2/2m$ multiplier (a large factor). Hence, it is common (but not necessary) to ignore this term in writing h_e. See Pack and Hirschfelder (1968) for further justification of this idea.

In seeking eigenstates ψ of $H = h_e + h_N$, it is usual to introduce the eigenfunctions of h_e as a basis for expressing the r_i dependence of ψ. What does this mean? Since h_e is a Hermitian operator in r_i space (which also contains reference to the locations of the nuclei), the eigenfunctions ϕ_k of h_e

$$h_e(r_i|R_a)\phi_k(r_i|R_a) = E_k(R_a)\phi_k(r_i|R_a) \tag{1.3}$$

form a complete set of functions of r_i. Note that h_e depends on R_a even though it is not an *operator* in the R_a space. Hence, the E_k and ϕ_k will vary as R_a varies. However, for any specific R_a, the set of $\{\phi_k\}$ is complete in r_i space. Hence, because it describes motion of electrons and nuclei, the *total* wave function ψ, which depends on r_i and R_a, can be expanded to yield

$$\psi(r_i, R_a) = \sum_k \phi_k(r_i|R_a)\chi_k(R_a) \tag{1.4}$$

in which, for now, the $\chi_k(R_a)$ can be viewed as "expansion coefficient functions" that are to be determined from the equation $H\psi = E\psi$.

Substituting the expression for ψ in equation 1.4 into the *total* Schrödinger equation, premultiplying by $\phi_l^*(r_i|R_a)$, and integrating over the *electronic* coordinates $\{r_j\}$, we obtain for a diatomic molecule (analysis of polyatomic molecules is more tedious but gives rise to no new features)

$$\sum_k \int (\phi_l^*[h_e\phi_k\chi_k - E\phi_k\chi_k - \frac{\hbar^2}{2\mu}(\phi_k\nabla_R^2\chi_k + \chi_k\nabla_R^2\phi_k \\ + 2\nabla_R\phi_k \cdot \nabla_R\chi_k)]dr_1 dr_2 \ldots dr_N = 0 \tag{1.5}$$

in which the symbol R is now used for the internuclear distance $R \equiv |\mathbf{R}_a - \mathbf{R}_b|$. Recall that the center-of-mass motion has been removed. Using equation 1.3 and the orthonormality of the $\{\phi_k\}$, this equation reduces to

$$\left[E_l - E - \frac{\hbar^2}{2\mu}\nabla_R^2\right]\chi_l = -\sum_k \left[\int \phi_l^* \nabla_R^2 \phi_k \, d\mathbf{r} \chi_k \frac{\hbar^2}{2\mu}\right.$$

$$\left. + 2\frac{\hbar^2}{2\mu}\int \phi_l^* \nabla_R \phi_k d\mathbf{r} \cdot \nabla_R \chi_k\right]. \quad (1.6)$$

The primary fact to notice in equation 1.6 is that there is *coupling* between the electronic states ϕ_k and ϕ_l caused by the fact that ϕ_k and ϕ_l depend upon $\mathbf{R}_a$ and, hence, vary as $\mathbf{R}_a$ moves. Thus, the ψ function of equation 1.4 cannot be expressed as a *single* product $\phi_k \chi_k$ but requires all of the *electronic* wavefunctions to describe even a single *total* state wavefunction. Faced with the problem that it is not possible to express the exact solution as an electronic wavefunction multiplied by a vibration/rotation function, an approximation is needed. Two approximations are described in the following section.

1.1. Born-Oppenheimer and Adiabatic Approximations

In the Born-Oppenheimer approximation, *all* of the terms on the right-hand side of equation 1.6 (including the $\phi_k = \phi_l$ term) are ignored. This procedure is equivalent to *assuming* that ψ can be approximated as $\phi_l \chi_l$ *and that* ϕ_l does not vary (strongly) with $\mathbf{R}_a$. Then, equation 1.6 is the Schrödinger equation for the motion (vibration/rotation) of the nuclei in the *potential energy field* $E_l(\mathbf{R}_a) = V_l$, namely,

$$\left(-\frac{\hbar^2}{2\mu}\nabla_R^2 + V_l\right)\chi_l = E\chi_l. \quad (1.7)$$

This equation states that the electronic energy, which certainly depends on where the nuclei are located, provides the potential energy surface on which the nuclei move. (Note that this potential surface is different for different electronic states labeled by l.) Thus, equation 1.7 is nothing but the vibration/rotation (V/R) problem and χ_l is one of the V/R wavefunctions for the lth state. In other words, $\chi_l = \chi_{l,v,J}$ and $E = E_{l,v,J}$; v and J are the vibration and rotation quantum numbers.

In the adiabatic approximation, the $k = l$ terms on the right-hand side of equation 1.6 are retained. As a result, the potential surface felt by the nuclei also includes the terms

$$\int \phi_l^* \left(-\frac{\hbar^2}{2\mu} \nabla_R^2 \phi_l\right) d\mathbf{r} \quad \text{and} \quad \int \phi_l^* \left(-\frac{\hbar^2}{\mu} \nabla_R \phi_l\right) d\mathbf{r} \cdot \nabla_R.$$

The V/R wavefunctions χ_l then also depend upon these "non-Born-Oppenheimer (BO) correction terms." Very few calculations in the literature have included such corrections. In the nonadiabatic approximation one attempts to keep all, or at least the most significant, terms in equation 1.6, but *ab initio* calculations at this level have been done only on very small systems (see Kolos and Wolniewicz, 1963, 1964, 1965).

Whenever we have two or more surfaces (E_1 and E_2) that come close together, we *must* consider the coupling of their electronic and nuclear motions. The usual way to think of this is to assume that the two *unperturbed* problems (ignoring the right-hand side of equation 1.6) for $\phi_1 \chi_{1,vJ}^0$ and $\phi_2 \chi_{2,v'J'}^0$ have been solved. Then we attempt to represent the true ψ as a combination of these two most important terms with unknown coefficients. The resulting 2×2 secular problem has diagonal elements $E_{1,vJ}^0$ and $E_{2,v'J'}^0$; for specific choices of vJ and $v'J'$ these elements can be nearly degenerate. The off-diagonal terms are

$$-\frac{\hbar^2}{2\mu} \langle \phi_1 \chi_{1,vJ}^0 | (\nabla_R^2 \phi_2) \chi_{2,v'J'}^0 + 2(\nabla_R \phi_2) \cdot \nabla_R \chi_{2,v'J'}^0 \rangle.$$

These non-Born-Oppenheimer coupling matrix elements, which determine the splitting between the two potential surfaces, will be large in regions of **R**-space in which the electronic wavefunctions are expected to undergo large changes in their bonding characteristics (for example, when changing from ionic to covalent bonds or when breaking old bonds and forming new bonds).

Although it might not be important to be able to perform quantitative *ab initio* quantum calculations that include non-Born-Oppenheimer terms, it is important to *understand* when such terms are likely to be large, because it is under these circumstances that the concept of the separate or uncoupled potential energy surfaces (V_l) breaks down. Alternatively, the idea of potential energy surfaces can be kept, and the coupling terms on the right-hand side of equation 1.6 can be viewed as giving rise to *transitions* from one surface to another. Such so-called *radiationless transitions* become important when the potential energy surfaces of the electronic states ϕ_k and ϕ_l approach or intersect one another. This problem of the rate of transitions among surfaces will be treated in more detail later (Chapters 5 and 6), when photochemical processes in which a molecule is prepared in an excited state are considered.

1.2. Intersections of Potential Energy Surfaces

In section 1.1 reference was made to electronic potential energy surfaces that intersect. Let us now briefly examine the circumstances in which two surfaces actually can "cross" one another. Consider a pair of approximate *electronic* wavefunctions ϕ_1 and ϕ_2 that might correspond to two different electronic configurations (e.g., Na^+, Cl^- and $Na\cdot$, $Cl\cdot$) of the same molecule. Alternatively, one could be referring to the coupling of zeroth-order Born-Oppenheimer wavefunctions to yield the full non-Born-Oppenheimer ψ, as discussed above. In the former case, the 2 × 2 secular problem that results from using these two functions as a basis for approximating the correct electronic wavefunctions ϕ_1 and ϕ_2 has energy levels given by the expression

$$E_\pm = \frac{1}{2}\left[h_{11} + h_{22} \pm \sqrt{(h_{11} - h_{22})^2 + 4h_{12}^2}\right] \quad (1.8)$$

in which

$$h_{ij} \equiv \int \phi_i^* h_e \phi_j \, d\mathbf{r}. \quad (1.9)$$

To make the two energy levels $E_\pm$ equal (for surface intersection) it is necessary that $h_{11} = h_{22}$ and $h_{12} = 0$ at the *same* geometrical point(s). For a diatomic molecule, the elements h_{ij} are functions of R only, so it is not generally possible to find R-values at which *both* of the above conditions are met. As a result, potential energy curves of diatomic molecules do not cross (unless ϕ_1 and ϕ_2 have different symmetry and h_{12} is identically zero for all R). For a general molecule with N atoms, there are $3N - 5$ (linear) or $3N - 6$ (nonlinear) vibrational degrees of freedom upon which $E_\pm$ can depend. By insisting that $h_{11} = h_{22}$ and $h_{12} = 0$, the dimension of the space in which $E_\pm$ can intersect is reduced to $3N - 7$ or $3N - 8$ (for two states of the same symmetry for a nonlinear molecule). Hence, states of the same symmetry *can* cross, though they cross on a surface whose dimension is two less than that of the potential energy surfaces on which the molecule is moving. As a result, the molecule does not frequently encounter such crossing geometry, so the fact that the surfaces may actually cross at special points is not particularly important. The essential point is that when surfaces approach one another closely (e.g., in the neighborhoods of crossings), transitions are likely to occur. (The rates of these transitions are discussed in Chapter 6.) The extension of the above analysis to intersections among more than two surfaces is nontrivial and has been given by Alden Mead (1979).

In summary, the potential energy surfaces upon which chemists usually base models for understanding the energetics of thermal and photochemical

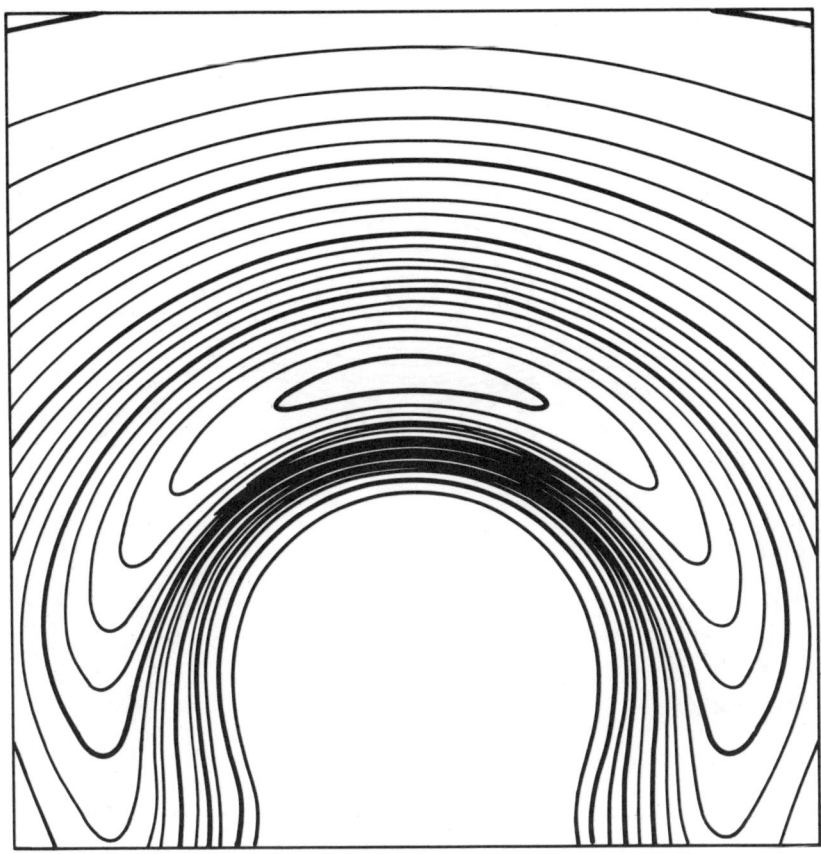

Figure 1-1
Potential energy surface of $X^1\Sigma$ ground-state HCN as a function of the location of the H atom for fixed CN bond length. The minimum on this surface is located at the linear ($\theta = 180°$) geometry with the H atom bonded to the C atom.

reactions can be thought of as solutions to the Born-Oppenheimer version of the *electronic* Schrödinger equation (equation 3). The dependence of these electronic energy levels $\{E_k\}$ on the internal coordinates of the molecule is what generates the potential surfaces that are depicted in many texts (for example, see Eyring, Walter, and Kimball, 1944; Pearson, 1976). This concept is illustrated in Figures 1-1 and 1-2 by contour graphs of the potential energy surfaces of the ground ($X^1\Sigma$) and $n\pi^*$ excited (C^1A') states of HCN as functions of the H—C bond length (r) and HCN bond angle (θ).

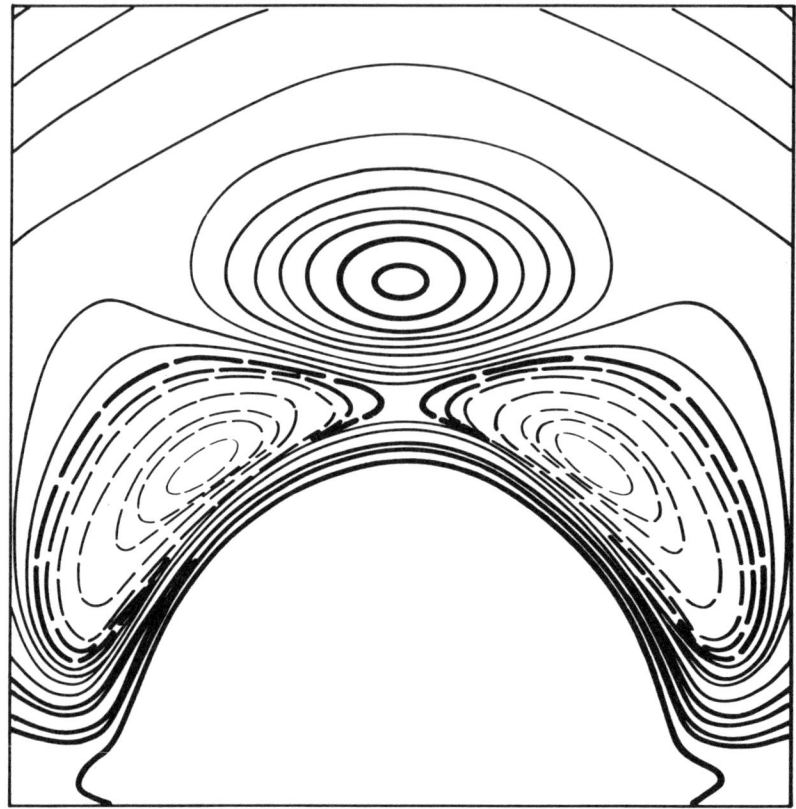

Figure 1-2
Potential energy surface of the C^1A' state of HCN as a function of the location of the H atom for fixed CN bond length. Note the potential well at an HCN angle near 140° and the barrier at 180°. Dashed contours are lower in energy than the solid contours.

Chapter 2

Symmetry and Potential Energy Surfaces

In this chapter two points are considered: (1) the information provided by the shape and topology of a surface and (2) how the shape of the surface makes the nuclei (molecular framework) move in a way that might lower the symmetry of the molecule.

In general, a potential energy surface is a function of $3N-5$ or $3N-6$ internal coordinates. For example, for HCN these coordinates could be r_{CH}, r_{CN}, and θ_{HCN}. At local minima on the energy surface small displacements of any of these internal coordinates $\{X_i\}$ increase the electronic energy. (Note that more than one minimum might be present, as, in the case of HCN and HNC.) Hence, at the local minima, the slopes or gradients vanish,

$$\left(\frac{\partial E}{\partial X_i}\right)_{min} = 0, \tag{2.1}$$

and the curvatures are positive, that is,

$$\left(\frac{\partial^2 E}{\partial X_i^2}\right)_{min} > 0, \tag{2.2}$$

and

$$\det\left(\frac{\partial^2 E}{\partial X_i \partial X_j}\right)_{min} > 0. \tag{2.3}$$

An alternative statement is that the gradients vanish and the eigenvalues of the Hessian matrix $(\partial^2 E/\partial X_i \partial X_j)$ are positive. Notice that it is possible that although equations 2.1-2.3 are obeyed, the potential well located at this minimum may not be deep enough to hold a bound vibrational state (if the zero-point vibrational energy is greater than the dissociation energy of the well).

At an activated complex or transition state, equation 2.1 is still valid for *all* coordinates, but along *one* special direction (which generally will be some

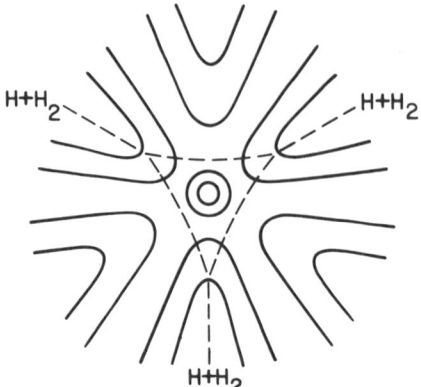

Figure 2-1
Schematic plot of potential surfaces for the exchange reaction $H + H_2 \to H_2 + H$. The central point is symmetric H_3, which is at a maximum in energy. The dashed lines are the reaction coordinates actually followed.

combination of all of the X_i—that is, $Q_r = \sum_i a_{ir} X_i$) the curvature is negative, $\partial^2 E/\partial Q^2 < 0$, with all other curvatures being positive. This direction is called the *reaction coordinate*. If this direction is followed away from the transition state, the slope $\partial E/\partial Q_r$ becomes nonzero. In particular, if we move along Q_r in a manner that maintains all other slopes at zero, namely $\partial E/\partial X_i = 0$ and $\partial^2 E/\partial X_i^2 > 0$, then it is said that one is "walking along a reaction path." There may, of course, be more than one reaction path and more than one transition state on the potential energy surface of a molecule, which simply means that there is more than one reaction event that this molecule (or "super" molecule) can undergo. By super molecule is meant the total system consisting of all atoms involved in the reaction. For example, H_2CO can undergo decomposition to yield either $H_2 + CO$ or $H + HCO$. Hence, H_2CO, $H_2 + CO$, and $H + HCO$ all consist of the same super molecule.

From the above discussion, we see that the reaction coordinate Q_r traces out the "valley floor," defining the reaction path that connects the reactants to the activated complex (where even $\partial E/\partial Q_r = 0$). Because a "mountain pass" such as that described above for the activated complex has only one direction of negative curvature, it cannot connect more than two valleys, and hence the reaction coordinate Q_r must be nondegenerate at the activated complex. Figure 2-1 illustrates this situation; it should be clear that the top of the mountain cannot be an activated complex. Lower-energy pathways exist for getting from one valley to another, and along these lower-energy paths there are nondegenerate motions with negative curvatures that have a point at which $\partial E/\partial Q_r = 0$.

It is *not* true that the coordinates of a molecule actually move *only* along the reaction coordinate in a real chemical reaction. Q_r is a characteristic of the potential surface; the internal motion of a molecule depends both upon the surface and upon how the collision occurs (initial molecular orientations, velocities, etc.). However, it is still useful to learn how to follow the reaction coordinate from local minima characterizing reactants through the activated complex to products. If this path can be followed, one might be able to predict whether the energy of the activated complex is high (i.e., if a large activation energy is expected). Hence, even though we are primarily interested in *energetics* (in contrast to the actual *dynamics* that the molecule undergoes on the surface), knowledge of the shape of the potential surface is important. To carry out *dynamic* studies of reaction mechanisms requires knowledge of the potential energy surface(s) at all geometries that are energetically accessible—not only along Q_r. Although such studies are becoming common in modern research in chemical dynamics, we shall focus on considerations of energetics and hence be satisfied to walk along or near Q_r.

From the slopes and curvatures of a potential energy surface at some starting geometry $\{\mathbf{R}_a^0\}$, automated algorithms (Cerjan and Miller, 1981; Simons, Jørgensen, Taylor, and Ozment, 1983) can be used to walk along Q_r from $\{\mathbf{R}_a^0\}$ to some new geometry $\{\mathbf{R}_a^1\}$. Continuing this step-by-step procedure, a transition state can eventually be reached. Further walking leads to the product state determined by this particular transition state. Such step-by-step walks are now routinely carried out in theoretical studies of chemical reactions. In this book, such walks will be conceptualized but not performed quantitatively.

The concepts just described relating to the shape of a potential energy surface can be used to determine whether an electronic wavefunction ϕ_0, which has energy E_0 at a starting geometry $\{Q_i^0\}$, corresponds to a local minimum (e.g., stable conformer), to a transition state, or to some point lying along the reaction coordinate. These conclusions can be made more quantitative. To do so, the dependence of the *electronic* Hamiltonian, which determines the energy surfaces, on the internal coordinates of the molecule must be examined.

The only term in $h_e(\mathbf{r}_i | \mathbf{R}_a)$ that is an electronic operator and depends on $\mathbf{R}_a$ is the electron-nuclear coulomb interaction

$$V_{\text{en}} \equiv -\sum_{i,a} Z_a e^2 |\mathbf{r}_i - \mathbf{R}_a|^{-1}.$$

Consider now how this interaction energy would change if some geometrical coordinates were changed by a small amount. The coordinate Q that is changed may be some combination of the x, y, and z position coordinates of each of the nuclei ($Q = \sum_a \mathbf{C}_a \cdot \mathbf{R}_a$). The change in V_{en} caused by a small change in Q can be expressed as

$$\frac{\partial V_{\text{en}}}{\partial Q} = \sum_a (\nabla_{\mathbf{R}_a} V_{\text{en}}) \cdot \left(\frac{\partial \mathbf{R}_a}{\partial Q}\right). \tag{2.4}$$

The derivative $\nabla_{\mathbf{R}_a} V_{en}$ is a vector containing the X_a, Y_a, and Z_a derivatives of V_{en}. For example,

$$\frac{\partial}{\partial X_a} V_{en} = -\sum_i Z_a e^2 |\mathbf{r}_i - \mathbf{R}_a|^{-3}(X_i - X_a), \tag{2.5}$$

in which X_i and X_a are the x coordinates of the ith electron and the ath nucleus, respectively, Notice that $\partial V_{en}/\partial X_a$ has X-symmetry as an operator in the space of the electrons, so $\nabla_{\mathbf{R}_a} V_{en}$ is a vector whose three components have x, y, and z symmetry as electonic operators. The $\partial \mathbf{R}_a/\partial Q$ term is nothing but the change in $\mathbf{R}_a$ accompanying a unit change in Q. The reaction coordinate can also be written as a linear combination of the elementary nuclear coordinate displacements. In fact, the elementary displacements, which are $3N - 6$ or $3N - 5$ in number, can be combined (see Wilson, Decius, and Cross, 1955) to give an equal number of symmetry coordinates (Q_i):

$$Q_i = \sum_j C_{ij} R_j, \tag{2.6}$$

in which the $\{R_j\}$ are the x, y, or z displacement coordinates. Conversely, it is also possible to express the displacements in terms of the symmetry coordinates, namely,

$$R_j = \sum_i (C^{-1})_{ji} Q_i. \tag{2.7}$$

As will be seen shortly, only distortions that are totally symmetric contribute to the *slope* of the potential energy surface when one is on the reaction coordinate. Hence, for motions along the (symmetric) reaction coordinate Q_r, the derivative term appearing in equation 2.4 can be related to the symmetry coefficients C_{ij},

$$\frac{\partial R_j}{\partial Q_r} = (C^{-1})_{jr}. \tag{2.8}$$

That is, $\partial R_j/\partial Q_r$ is merely the element of the inverse transformation matrix. For example, in H_2O there are two stretching coordinates

$$Q_\pm = \frac{1}{\sqrt{2}} (\Delta r_{OH_1} \pm \Delta r_{OH_2}). \tag{2.9}$$

The inverse transformation is

$$\Delta r_{OH_{1,2}} = \frac{1}{\sqrt{2}}(Q_+ \pm Q_-). \tag{2.10}$$

Hence,

$$\frac{\partial \Delta r_{OH_{1,2}}}{\partial Q_+} = \frac{1}{\sqrt{2}}; \tag{2.11}$$

since Q_+ is the reaction coordinate (at the equilibrium geometry of H_2O), these derivatives are the values needed in equation 2.4. For the treatment of more complicated symmetry coordinates, see Wilson, Decius, and Cross (1955).

An important point is that $\partial V_{en}/\partial Q$ has the same symmetry (when considered to be an *electronic* operator) as Q itself has (when considered as a function of nuclear positions). How is this fact useful? Let us assume that we have an electronic wavefunction, which from now on we denote as ψ_0, that obeys the relation

$$h_e(r_i|\mathbf{R}_a^0)\psi_0 = E_0(\mathbf{R}_a^0)\psi_0 \tag{2.12}$$

(i.e., the electronic Schrödinger equation at a geometry $\mathbf{R}_a^0 \equiv Q^0$). *Perturbation theory* will first be used to compute the change in the electronic energy E_0 that accompanies a small change in some coordinate Q. The perturbation is the change in h_e brought about by a small movement in the Q direction

$$h_e(r_i|Q) = h_e(r_i|Q^0) + (\partial h_e/\partial Q)\Delta Q + \frac{1}{2}(\partial^2 h_e/\partial Q^2)\Delta Q^2 + \cdots$$

$$\equiv h_e^0 + V. \tag{2.13}$$

The change in the electronic energy E_0 can be expressed (through second order in ΔQ) using conventional perturbation theory (Eyring, Walter, and Kimball, 1944) as

$$E_0(Q) = E_0(Q^0) + \langle\psi_0|V|\psi_0\rangle_{Q^0} + \sum_{k\neq 0} \frac{|\langle\psi_k|V|\psi_0\rangle_{Q^0}|^2}{E_0(Q^0) - E_k(Q^0)}, \tag{2.14}$$

in which the ψ_k are the other eigenfunctions of h_e at Q^0.

2.1 Slope of the Energy Surface: First-Order Jahn-Teller Effect

Clearly the only term that is linear in ΔQ appears in the $\langle\psi_0|V|\psi_0\rangle_{Q^0}$ factor and gives our approximation to the *slope* of the potential surface along the Q direction

$$(\partial E_0/\partial Q)_{Q^0} = \langle\psi_0|\partial h_e/\partial Q|\psi_0\rangle_{Q^0}$$

$$= \sum_a (\partial \mathbf{R}_a/\partial Q)\cdot\langle\psi_0|-\sum_i Z_a e^2/|\mathbf{r}_i - \mathbf{R}_a|^3(\mathbf{r}_i - \mathbf{R}_a)|\psi_0\rangle$$

$$+ \sum_a (\partial \mathbf{R}_a/\partial Q)\cdot\sum_b \frac{Z_a Z_b e^2}{|\mathbf{R}_b - \mathbf{R}_a|^3}(\mathbf{R}_b - \mathbf{R}_a) \qquad (2.15)$$

The second term in equation 2.15 comes from taking the derivative with respect to Q of the nuclear-nuclear coulomb repulsion terms (V_{nn}) that were also included in h_e (these terms are not functions of the electronic coordinates). Because V_{nn} is a totally symmetric function of the nuclear positions (i.e., it displays the symmetry of the nuclear framework), any distortion Q that is not totally symmetric yields $\partial V_{nn}/\partial Q = 0$. For example, the antisymmetric stretching coordinate of H_2O does not change V_{nn}, since it moves one H atom closer to the O while the other moves farther away (the H—H distance remains constant). Thus, V_{nn} contributes to the slope of the surface only for totally symmetric distortions.

What about the symmetry effects in V_{en}? We saw earlier that $\partial V_{en}/\partial Q$ has the same symmetry as Q. If ψ_0 is nondegenerate (i.e., not *symmetry degenerate*), the product $\psi_0^*\psi_0$ is totally symmetric, and thus the integral $\langle\psi_0|\partial V_{en}/\partial Q|\psi_0\rangle$ will vanish unless $\partial V_{en}/\partial Q$, and hence Q, is also totally symmetric. Therefore, for nondegenerate states only totally symmetric distortions contribute to the slope of the potential surface; other kinds of motion automatically yield $\partial V_{nn}/\partial Q = 0$ and $\langle\psi_0|\partial V_{en}/\partial Q|\psi_0\rangle = 0$. At a minimum or at a saddle point, even the symmetric distortions give a total slope of zero, since

$$\partial V_{nn}/\partial Q = -\langle\psi_0|\partial V_{en}/\partial Q|\psi_0\rangle$$

at these special points. From the definition of Q_r, the reaction coordinate has to be totally symmetric *if* ψ_0 is nondegenerate, because the slope of the potential surface along Q_r is assumed to be nonzero except at local minima or at activated complexes.

What if ψ_0 is degenerate? In this case, the symmetry of $\psi_0^*\psi_0$ contains at least one element that is not totally symmetric and that itself may or may not be degenerate. [This result is a simple and well-known result that is treated in many texts on group theory (Cotton, 1963; Herzberg, 1966; Wigner, 1959)]. Thus, motion along nonsymmetric directions contributing to the slope through

$\langle\psi_0|\partial V_{en}/\partial Q|\psi_0\rangle$ (but not to $\partial V_{nn}/\partial Q$) is also possible. Moreover, motion along such nonsymmetric directions (in one or the other ± sense) will *lower* the electronic energy, so such degenerate points on the surface are generally not activated complexes (since the slope is nonzero at degenerate points). Hence we conclude that a degenerate state will generally be unstable to distortion along a nonsymmetric direction (which thereby lowers the overall symmetry of the molecule). If this nonsymmetric distortion is itself degenerate then, though some of the original symmetry of the molecule may be lost, it is possible that not all of the symmetry is broken (by these linear $\langle\psi_0|\partial V_{en}/\partial Q|\psi_0\rangle$ terms). Molecules for which these slope terms are nonvanishing for degenerate states are said to be unstable with respect to first-order Jahn-Teller (FOJT) distortion. Of course, the symmetry of $\psi_0^*\psi_0$ also contains A_1 (the totally symmetric element), so symmetric distortions also give rise to nonvanishing slopes for degenerate states. However, such symmetric distortions will generally preserve the degeneracy of the state ψ_0.

At this stage of the analysis of movement along the reaction coordinate, the following points have been established about the potential surface: (1) At the activated complex, Q_r cannot be degenerate because a mountain pass can connect only two valleys; that is, the surface can have only one direction of negative curvature. (2) If ψ_0 is nondegenerate, Q_r must be totally symmetric. (3) If a point is reached at which ψ_0 is degenerate, a nonsymmetric motion will distort the molecule, thereby lowering its energy (remaining in the valley) and lowering its symmetry (so this motion is now symmetric in this lower-symmetry point group). This behavior of the reaction coordinate—that is, totally symmetric—makes the symmetry of ψ_0 unchanged (except when ψ_0 is degenerate) and leads to the concept of connecting states by symmetry (symmetry conservation).

2.2 Surface Curvature: Second-Order Jahn-Teller Effect

The effects of terms that determine the *slope* of the potential energy surface have just been described; now, curvature terms—those quadratic in Q_r—will be examined.

The quadratic terms are mainly of concern in regions of the potential surface at which the slopes are zero but at which the system might be unstable because it is at a saddle point—for example, at the activated complex. Equation 2.8 has two such terms. The first term

$$\langle\psi_0|\partial^2 V/\partial Q_r^2|\psi_0\rangle Q_r^0$$

concerns the response of the "frozen" charge density of ψ_0 to a change in $V_{nn} + V_{en}$. Q_r is totally symmetric except when ψ_0 is degenerate (in which case a nonsymmetric Q_r will distort the molecule and become symmetric in the lower symmetry), and $\partial^2 V/\partial Q_r^2$ has the same symmetry as Q_r^2. Hence, the term $\langle\psi_0|\partial^2 V/\partial Q_r^2|\psi_0\rangle$ is generally nonzero; in fact, it is also positive. This can be seen by evaluating $\partial^2 V/\partial Q^2$:

$$\frac{\partial^2 V}{\partial Q^2} = \sum_{a,b} \frac{\partial \mathbf{R}_a}{\partial Q} \cdot \frac{\partial^2 V}{\partial \mathbf{R}_a \partial \mathbf{R}_b} \cdot \frac{\partial \mathbf{R}_b}{\partial Q}. \tag{2.16}$$

Recall that V contains V_{nn} and V_{en} terms. Because

$$d^2/dx^2 |x-y|^{-1} = -4\pi\delta(x-y),$$

(see page 69 of Arfken, 1970) the expression can be evaluated. For V_{nn}, note that $Z_a Z_b |\mathbf{R}_a - \mathbf{R}_b|^{-1}$ contains $\mathbf{R}_a - \mathbf{R}_b$ in a symmetrical fashion; thus,

$$\partial/\partial\mathbf{R}_b |\mathbf{R}_a - \mathbf{R}_b|^{-1} = -\partial/\partial\mathbf{R}_a |\mathbf{R}_a - \mathbf{R}_b|^{-1}.$$

Therefore,

$$\frac{\partial^2 V_{nn}}{\partial \mathbf{R}_a \partial \mathbf{R}_a} = -4\pi \sum_{b \neq a} Z_a Z_b e^2 \delta(\mathbf{R}_a - \mathbf{R}_b) \tag{2.17}$$

and

$$\frac{\partial V_{nn}}{\partial \mathbf{R}_a \partial \mathbf{R}_b} = 4\pi Z_a Z_b e^2 \delta(\mathbf{R}_a - \mathbf{R}_b), \text{ for } a \neq b. \tag{2.18}$$

Since the nuclei never are located at the same position, these δ functions vanish.

Using the above δ-function identity, $\partial^2 V_{en}/\partial Q^2$ can be evaluated as

$$\frac{\partial V_{en}}{\partial \mathbf{R}_a \partial \mathbf{R}_a} = \sum_i (-Z_a e^2)(-4\pi)\delta(\mathbf{r}_i - \mathbf{R}_a) \text{ and } \frac{\partial^2 V_{en}}{\partial \mathbf{R}_a \partial \mathbf{R}_b} = 0.$$

Thus,

$$\frac{\partial^2 V_{en}}{\partial Q^2} = \sum_i \sum_a (\partial \mathbf{R}_a/\partial Q)^2 4\pi e^2 Z_a \delta(\mathbf{r}_i - \mathbf{R}_a). \tag{2.19}$$

The expectation value of this term gives the first term in the curvature, namely,

$$\psi_0|(\partial^2 V_{en}/\partial Q^2|\psi_0 = \sum_a 4\pi Z_a e^2 (\partial \mathbf{R}_a/\partial Q)^2 \rho(\mathbf{R}_a) \tag{2.20}$$

in which $\rho(\mathbf{R}_a)$ is the electron density in state ψ_0 at the nucleus at $\mathbf{R}_a$. Clearly, this contribution to the curvature is always positive and will be nonzero for any symmetry of ψ_0, since $\partial^2 V/\partial Q^2$ is totally symmetric. The negative curvature of the surface at an activated complex is a result of a second contribution to the curvature. This is given by

$$\sum_{k\neq 0} |\langle\psi_k|(\partial V/\partial Q)|\psi_0\rangle|^2 (E_0 - E_k)^{-1}$$

and is always negative (if ψ_0 is the ground state) because $E_0 - E_k$ is negative. Earlier, it was shown that $\partial V/\partial Q$ has the same symmetry as Q. Therefore, if Q is totally symmetric (as it is along the reaction coordinate where ψ_0 is nondegenerate), the excited state ψ_k must have the same symmetry as ψ_0. On the other hand, if Q is not symmetric, which might occur at a minimum or maximum point at which *all* $\partial E/\partial Q = 0$ (and hence consideration of the quadratic terms in $E_0(Q)$ becomes essential), or if ψ_0 were degenerate, so that Q leads to distortion of the molecule, then the symmetry of ψ_k is dictated by the direct product of the Q and ψ_0 symmetries. Notice that because the slope of E_0 is zero at the activated complex, the energy variation is now dictated by the quadratic terms, which can *now* allow Q_r to be nonsymmetric.

Clearly, for these negative curvature terms to become important (and even dominant, as they are at an activated complex), the symmetry of ψ_k must be correct *and* the energy splitting $E_0 - E_k$ must be small. This situation occurs when a chemical bond is broken. For example, at large internuclear distances the σ^2 and $\sigma^1\sigma^{*1}$ configurations of HCl are reasonably close together in energy. Because $\partial V/\partial Q$ is a one-electron operator, the excited states ψ_k that can couple most strongly with ψ_0 are those that are singly excited relative to ψ_0 (Condon and Shortley, 1957; Cook, 1978). As a result, negative curvature along the reaction path should be possible when there are low-lying excited states that involve single promotions of electrons from bonding orbitals in ψ_0 to antibonding orbitals in ψ_k.

To gain more insight into why ψ_k and ψ_0 should be related in this antibonding/bonding manner, recall that we are looking (using perturbation theory) at the response of the system (ψ_0, E_0) to a small displacement of the nuclei (Jørgensen and Simons, 1981). The energy response has already been discussed above. The change in the wavefunction caused by the perturbation V is given by

$$\psi_0 \to \psi_0 + \sum_{k\neq 0} \langle\psi_k|V|\psi_0\rangle (E_0 - E_k)^{-1} \psi_k \tag{2.21}$$

(Eyring, Walter, and Kimball, 1944). Thus, the electron density $\psi_0^*\psi_0$ changes (through first order in the change in Q) by an amount

$$\sum_{k\neq 0} 2\psi_0^*\psi_k \langle\psi_k|\partial V/\partial Q|\psi_0\rangle (E_0 - E_k)^{-1} \equiv \sum_{k\neq 0} \delta\rho_{0k}. \qquad (2.22)$$

Where $\delta\rho_{0k}$ is positive, the electron density increases as the motion along Q occurs; where it is negative, electron density decreases. The symmetry (i.e., the nodal pattern) of $\delta\rho_{0k}$ can be determined by looking at the symmetry of $\psi_0^*\psi_k$. If ψ_0 and ψ_k are approximated by Slater determinants (Cook, 1978) that differ by a single orbital replacement ($\phi_0 \to \phi_k$), the nodal pattern is that of the orbital products $\phi_0^*\phi_k$. The positive nuclei will move to regions at which $\delta\rho_{0k}$ is positive (i.e., in which electron density piles up) and will leave regions in which $\delta\rho_{0k}$ is negative.

Consider, for example, the H$_2$O molecule at its equilibrium geometry. Since ψ_0 is nondegenerate, all of the slope terms vanish. What about the curvatures? Excitation of an electron from the bonding a_1 OH orbital to its antibonding a_1 partner gives a $\phi_0^*\phi_k$ pattern of the form

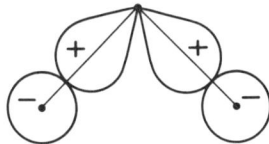

which is consistent (according to the above analysis of the integrals arising in the curvature terms) with a symmetric stretch distortion. The bonding b_2 to antibonding $b_2\delta\rho_{0k}$ also looks like

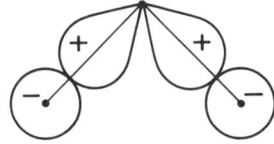

which is also consistent with a symmetric stretch. On the other hand, the $a_1 \to b_2$ or $b_2 \to a_1$ excitations have a $\delta\rho_{0k}$ of the form

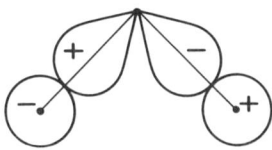

which is consistent with an antisymmetric b_2 stretch. Of course, we do not expect any of these excitations to give rise to large (negative) contributions to the curvatures in this particular (H$_2$O) case. Their excitation energies, which occur in the denominators of the curvature terms, are very large, since they involve $\sigma_{OH} \to \sigma_{OH}^*$ excitations. As a result, the negative contributions made by these excitations are more than outweighed by the positive contributions arising from the terms shown in equation 2.20. This is, of course, expected here, since we are considering H_2O at its equilibrium geometry and negative curvature is not anticipated.

Before proceeding to the application of the ideas presented in the first two chapters, it is useful to review the facts that have been established about the reaction coordinate, the activated complex, and the slope and curvature of the surface along Q_r. Remember that the goal is to be able to use this information to move along Q_r from reactions, through an activated complex, to products in order to estimate the activation energy for a reaction. As this path is taken, the symmetry of the wavefunction remains conserved except when the state becomes degenerate (first-order Jahn-Teller) or when low-lying singly excited states come into play and give rise to second-order Jahn-Teller distortions.

Chapter 3

Review of Molecular Orbital and Configuration-Mixing Ideas

Chapters 1 and 2 describe what a potential surface is and how a reaction path moves along the surface. In this chapter we discuss briefly how to obtain such surfaces. Knowledge of the procedure is important because it relates to the actual *numerical* evaluation of potential energy surfaces and, furthermore, because it makes one think about those electronic configurations that are likely to be important in describing chemical reactions. These ideas are presented more thoroughly in Appendix A.

Consider again the hypothetical dissociation of H_2O. At its equilibrium geometry this molecule has molecular orbitals with a_1, b_2, and b_1 symmetries. The $1s_O$, σ_{OH}, σ_{OH}^*, and lone pair orbitals in the molecular plane all have a_1 symmetry. Another σ_{OH} and σ_{OH}^* orbital pair has b_2 symmetry, and the p_0 orbital directed perpendicular to the molecular plane has b_1 symmetry. As we saw above, the $\sigma_{OH} \to \sigma_{OH}^*$, $a_1 \to b_2$, or $b_2 \to a_1$ orbital excitations of H_2O may play important roles in the asymmetric dissociation to give $OH + H$. Hence, we expect that the $1a_1^2 2a_1^2 1b_2^2 3a_1^2 1b_1^2$, $1a_1^2 2a_1^2 1b_2 3a_1^2 1b_1^2 4a_1$, and $1a_1^2 2a_1 1b_2^2 3a_1^2 1b_1^2 2b_2$ configurations should be important in describing this fragmentation. Although the orbitals having a_1 and b_2 symmetry can only be labeled as a' once the C_{2v} symmetry is broken (b_1 becomes a''), we can immediately tell that the first configuration above cannot possibly describe $OH + H$ because all orbitals are doubly occupied, whereas the radical fragments $OH + H$ have two singly occupied orbitals. The other two configurations do have the correct orbital occupancy to describe $OH + H$. However, at the equilibrium geometry of H_2O, this first configuration dominates the electronic wavefunction because it has two pairs of bonding electrons. Hence, as H_2O fragments, a substantial change in the electronic structure is expected to occur when moving from one dominant configuration to another.

Before proceeding further to specific examples, one must learn how to construct wavefunctions whose energies give us the desired potential energy surfaces. In the conventional molecular orbital model of electronic structure (Cook, 1978; Pilar, 1968) there are three levels of analysis of wavefunctions:

the orbital, configuration, and state-function levels. The electronic wavefunction of a given state is usually expressed as a linear combination of configurations, each of which is expressed in terms of Slater determinants over molecular orbitals (see Appendix A). In some circumstances, the state wavefunctions of the reactants and products may be smoothly connected (correlated) by way of the reaction coordinate, though the orbitals or orbital occupancies (configurations) of reactants and products may not correlate smoothly. Recall that we are directed to correlate or to connect orbitals, configurations, or states by the observation that Q_r is totally symmetric (except where ψ_0 is degenerate or when second-order Jahn-Teller effects dominate), and hence, movements along Q_r cannot change the symmetry of ψ_0.

3.1 Molecular Orbitals: Symmetry of the Fock Operator

Let us recall from Chapter 1 how Hartree-Fock (HF) molecular orbitals, which are probably the most widely used orbitals, are obtained. (For those readers who wish to review the fundamental steps involved in *ab initio* molecular orbital calculations, a brief overview is provided in Appendix A.) A Fock operator can be constructed from a particular orbital occupancy that is assumed to dominate the true wavefunction at the geometry at which one is located [we now write the operators in atomic units as in Pilar (1968)]:

$$F = -\frac{1}{2}\nabla_r^2 - \sum_a \frac{Z_a}{|\mathbf{r}-\mathbf{R}_a|} + \sum_\mu \int \phi_\mu^*(\mathbf{r}_2) \frac{1-P_{12}}{|\mathbf{r}-\mathbf{r}_2|} \phi_\mu(\mathbf{r}_2)\, d\mathbf{r}_2 \qquad (3.1)$$

in which μ extends over all of the spin orbitals that appear in the presumed dominant electronic configuration. Clearly, the first two terms in F commute with the symmetry operations of the molecule because they depend on $\mathbf{R}_a$ in a symmetrical manner. If the ϕ_μ are nondegenerate and symmetry-adapted (this will often be true in so-called symmetry-restricted HF calculations), $\phi_\mu^*(\mathbf{r}_2)\phi_\mu(\mathbf{r}_2)$ is totally symmetric: therefore, even the coulomb part of the last term in F will commute with all symmetry operations. To show that the exchange part is also symmetric is more difficult.

Consider the commutator of the exchange operator K with any symmetry operation σ

$$[\sigma, K]\phi(\mathbf{r}) = \sigma(\mathbf{r}) \int K(\mathbf{r}, \mathbf{r}_2)\phi(\mathbf{r}_2)\, d\mathbf{r}_2 - \int K(\mathbf{r}, \mathbf{r}_2)\sigma(\mathbf{r}_2)\phi(\mathbf{r}_2)\, d\mathbf{r}_2 \qquad (3.2)$$

in which the kernel $K(\mathbf{r}, \mathbf{r}_2)$ is defined as

$$\int K(\mathbf{r}, \mathbf{r}_2)\phi(\mathbf{r}_2)\, d\mathbf{r}_2 = \sum_\mu \int \phi_\mu^*(\mathbf{r}_2) \frac{1}{|\mathbf{r}-\mathbf{r}_2|} \phi(\mathbf{r}_2)\, d\mathbf{r}_2 \phi_\mu(\mathbf{r}). \qquad (3.3)$$

Now using the fact that symmetry operators are unitary ($\sigma^+ = \sigma^{-1}$), we obtain

$$\int K(\mathbf{r}, \mathbf{r}_2)\sigma(\mathbf{r}_2)\phi(\mathbf{r}_2)\, d\mathbf{r}_2 = \int [\sigma^{-1}(\mathbf{r}_2)K(\mathbf{r}, \mathbf{r}_2)]\phi(\mathbf{r}_2)\, d\mathbf{r}_2 \qquad (3.4)$$

and hence

$$[\sigma, K]\phi(\mathbf{r}) = \int [\sigma(\mathbf{r}) - \sigma^{-1}(\mathbf{r}_2)]K(\mathbf{r}, \mathbf{r}_2)\phi(\mathbf{r}_2)\, d\mathbf{r}_2. \qquad (3.5)$$

From equation 3.3 it should be clear that $K(\mathbf{r}, \mathbf{r}_2)$ contains $\mathbf{r}$ and $\mathbf{r}_2$ in a symmetrical manner. Moreover, for abelian point groups (those with no degenerate representations; see Cotton, 1963), $\sigma^{-1} = \sigma$. Therefore, $\sigma(\mathbf{r}) - \sigma^{-1}(\mathbf{r}_2)$ operating on $K(\mathbf{r}, \mathbf{r}_2)$ would give zero, and the commutator $[\sigma, K]$ vanishes. For nonabelian groups σ^{-1} is no longer σ. However, if the sum over occupied spin orbitals $\{\sigma_\mu\}$ has equal occupancy for sets of (degenerate) orbitals that are related to one another by symmetry (i.e., $\sigma\phi_\mu = \phi_{\mu'}$), then the overall sum arising in $\sigma(\mathbf{r})K(\mathbf{r}, \mathbf{r}_2)$ will be the same (although not term-by-term) as that in $\sigma^{-1}(\mathbf{r}_2)K(\mathbf{r}, \mathbf{r}_2)$ and again $[\sigma, K] = 0$. The main point is that $[\sigma, F] = 0$ implies that the eigenfunctions of F, which are the Hartree-Fock molecular orbitals, will also be eigenfunctions of σ and, hence, will be symmetry adapted. As a result, all of the rules for correlating *states* (ψ_0, ψ_k) that are discussed above immediately apply also to these Hartree-Fock *orbitals* since F has all of the same symmetry as h_e. This means that symmetry conservation applies to orbitals *and* to total wavefunctions.

Now let us review how the Hartree-Fock equations are solved for the molecular orbitals. First, an atomic-orbital basis set consisting most likely of orbitals of the Slater ($r^{n-1}e^{-\zeta r}Y_{lm}$) or the Gaussian ($x^a y^b z^c e^{-\alpha r^2}$) type is chosen. These basis functions generally are located on each of the nuclei in the molecule being studied. Minimal, double-zeta, or extended bases including polarization functions are common choices. Tabulations of good basis sets are available for the ground-state normal chemical-valence states of most first- and second-row atoms as they occur in molecules. For example, good Gaussian bases are given by Huzinaga (1965) and by Dunning (1970, 1971). If the state of interest has unusual behavior (i.e., ionic states, Rydberg states, or many low-lying excited states), it is necessary to explore the effect of adding more and more atomic basis functions. The importance of this basis-set selection step cannot be overemphasized; without a good basis, one has little chance of achieving meaningful results.

Once an atomic basis is obtained, all one-electron (kinetic energy, overlap, and electron-nuclear interactions) and two electron ($\langle ab|cd \rangle$) integrals are evaluated (with a computer). At this stage in the calculation, most computer programs make use of symmetry information to combine basis functions into symmetry-adapted functions $\{\chi_b\}$ and to generate the one- and two-electron integrals over these symmetry functions.

The matrix elements of the Fock operator are then constructed (Roothaan, 1951) within the symmetry-adapted basis. This is done symmetry-by-symmetry since **F** is block-diagonal. To construct each block of **F** *all* (i.e., those belonging to all symmetries) of the *occupied* spin orbitals $\{\phi_\mu\}$ must be available. However, these Hartree-Fock molecular orbitals are not yet known, so an iteration process is used (Cook, 1978). With the aid of a computer one can guess the form of the occupied molecular orbitals; this is done by specifying the expansion coefficients $\{\mathbf{C}_{\mu b}\}$ of ϕ_μ in the symmetry-adapted basis:

$$\phi_\mu = \sum_b \chi_b \mathbf{C}_{\mu b}. \tag{3.6}$$

The guess can be made either on chemical grounds (e.g., $\phi_1 = 1s_O$ for H_2O) or, as in most computer programs, by first solving the equation

$$F\phi_\mu = \epsilon_\mu \phi_\mu \tag{3.7}$$

ignoring the coulomb and exchange contributions to F. The orbitals that result from the latter procedure are usually not chemically reasonable because they respond to only the isolated nuclei—no electron repulsion (screening) effects were included. Nevertheless, these initial orbitals can be used to construct a new F operator whose matrix elements (in the symmetry-adapted basis) are defined by

$$F_{cb} = \left\langle \chi_c \left| -\frac{1}{2}\nabla_r^2 - \sum_a \frac{Z_a}{|\mathbf{r}-\mathbf{R}_a|} \right| \chi_b \right\rangle$$

$$+ \sum_\mu \sum_{d,f} C_{\mu d} C_{\mu f} [\langle \chi_c \chi_d | \chi_b \chi_f \rangle - \langle \chi_c \chi_d | \chi_f \chi_b \rangle \delta_{s_\mu, s_c}], \tag{3.8}$$

(Roothaan, 1951) in which μ runs over the occupied *spin* orbitals and δ_{s_μ, s_c} indicates that the spin (α, β) of ϕ_μ must match that of χ_c for the exchange term to contribute. The form of the Fock matrix given in equation 3.8 is appropriate for performing a spin-unrestricted Hartree-Fock (UHF) calculation. There are two different **F** matrices for the α and β spin orbitals. Therefore, the molecular orbitals computed for α and β spin generally differ. Numerous techniques exist that attempt to overcome this somewhat inconvenient fact (different orbitals for different spins), so a single Fock matrix can be used to generate spatial orbitals that are appropriate for both α and β spins. We will not go further into these procedures here because we wish to emphasize the physical nature of the Hartree-Fock orbitals and equations rather than details of their solutions for special cases. These are treated in a clear manner by Cook (1978).

Having formed **F** (using the crude $\{C_{\mu,d}\}$), we solve

$$\mathbf{F}\mathbf{C}_\mu = \epsilon_\mu \mathbf{S}\mathbf{C}_\mu. \tag{3.9}$$

in which **S** is the overlap matrix, for a new set of $\{C_\mu\}$ coefficients that are then used to form a new **F** and, subsequently, a new set $\{C_\mu\}$. This iterative self-consistent-field (SCF) procedure is continued until the $\{C_\mu\}$ no longer vary from iteration to iteration.

The results of such an SCF calculation are a set of occupied and unoccupied (virtual) orbitals $\{\phi_\mu\}$ and orbital energies $\{\epsilon_\mu\}$. For example, for a double-zeta basis of H_2O, there are fourteen χ_c functions (eight s and six p). Hence, **F** is a 14×14 matrix having fourteen eigenvalues and fourteen eigenvectors. Of the fourteen SCF orbitals, only five are occupied in the ground state ($1a_1^2 2a_1^2 1b_2^2 3a_1^2 1b_1^2$); nine are virtual or unoccupied orbitals. Keep in mind that the words *occupied* and *virtual* only refer to the occupancy which *you* guessed to start the SCF procedure. We saw earlier that as H_2O is pulled apart to give $OH + H$, the occupancy changes. Thus, for $OH + H$ it would be more natural to use the "open shell" configuration to define occupancy.

3.2. When Can Orbital Energies be Added?

Before closing this discussion of orbitals, let us review (Cook, 1978; Pilar, 1968) the expression for the total electronic energy E_{HF} in the Hartree-Fock approximation:

$$E_{HF} = \sum_\mu \epsilon_\mu - \frac{1}{2} \sum_{\mu,\nu} \langle \mu\nu | \widetilde{\mu\nu} \rangle, \tag{3.10}$$

in which μ and ν run over the occupied spin orbitals and $\langle \mu\nu | \widetilde{\mu\nu} \rangle$ represents the coulomb interaction integrals minus the exchange integrals (Cook, 1978) over the Hartree-Fock molecular orbitals. It is important to note that the sum of the occupied orbital energies does *not* give E_{HF}, because, through F, each ϵ_μ contains interactions between ϕ_μ and *all* other ϕ_ν orbitals. Hence, the sum $\Sigma_\mu \epsilon_\mu$ doubly counts the electron-electron interactions. As a result, the second term in equation 3.10 is needed. Although $E_{HF} + \frac{1}{2} \Sigma_{a,b} (Z_a Z_b/R_{ab})$ is not equal to the sum of orbital energies plus the nuclear repulsion energies, the *changes* in this energy accompanying molecular distortion can, for *neutral* molecules, often be approximated well by the changes in $\Sigma_\mu \epsilon_\mu$. This approximation works because the change in $1/2 \Sigma_{a,b} (Z_a Z_b/R_{ab})$ is nearly perfectly cancelled by changes in $-1/2 \Sigma_{\mu,\nu} \langle \mu\nu | \widetilde{\mu\nu} \rangle$ arising from ϕ_μ and ϕ_ν that have major amplitudes on different centers. That is, the *subtracted* electron-electron repulsions involving orbitals on different atoms cancel the repulsions of the corresponding nuclei (at least at large bond length). This cancellation does not occur for ions because there are "extra" or "missing" electrons whose repulsions are not cancelled. One more thing must be stressed at this time: even though the shape of the HF-level potential energy surface might be well represented by the shape of $\Sigma_\mu \epsilon_\mu$, the entire Hartree-Fock picture rests on a guess of *the* dominant electronic configuration occupancy *and* the assumption that ψ_0 and E_0 could be accurately represented by a single determinant wavefunction. If the guess is wrong, or if the correct electronic wavefunction requires more than one configuration to describe reality qualitatively (e.g., in H_2O as it fragments into $OH + H$), the shape of the Hartree-Fock surface will probably not be correct.

3.3. Configuration Construction and Mixing

In the preceding sections the means by which molecular orbitals are defined, calculated, and correlated by symmetry along the reaction coordinate have been described. This information is not, however, sufficient to allow a statement about how the *wavefunctions* are to be symmetry-correlated—other information is needed about how the orbitals are occupied in the state wavefunction ψ_0. This amounts to specifying the electronic configurations that are important in describing ψ_0 throughout the entire range of the reaction coordinate. Many sophisticated *ab initio* computer programs (Shavitt, 1978) have configuration-selection subroutines that choose those configurations of the proper symmetry whose energies (expectation values) are low in order to represent the ground or low-lying excited states accurately.

In most chemical reactions, by using information about the orbital energy variations and estimates of electron repulsion energies, we can guess those few configurations likely to dominate ψ_0. For the asymmetric fragmentation of H_2O, we expect both the $(1-4)(a')^2(1a'')^2$ and the $(3a')^2 \to 3a'5a'$ configurations to be important. The former configuration dominates ψ_0 near the equilibrium geometry of H_2O, whereas the latter dominates for $OH + H$. For the $OH + H$ geometry the $(1-4)(a')^2(1a'')^2$ configuration corresponds to $OH^- + H^+$. At the equilibrium geometry of H_2O, the $(3a')^2 \to 3a'5a'$ configuration describes a singly excited state of H_2O that has one OH bond broken (i.e., $\sigma_{OH}^2 \sigma_{OH'} \sigma_{OH'}^*$).

In general, we first consider how the orbitals of the reactants and products symmetry-correlate along the reaction coordinate. This is done by simply ordering the orbitals of reactants and products by their energies and by connecting the orbitals of the same symmetry by "correlation lines." Then we attempt to write down all occupancies (configurations) of these orbitals that are consistent with the total state symmetry and that are expected to have low total electronic energy. From this list of dominant configurations, a qualitative diagram can be drawn displaying their energies (expectation values) as functions of the reaction coordinate. This diagram is referred to as a configuration-correlation diagram (CCD); it is the configuration-space analog of the orbital-correlation diagram (OCD).

The step of constructing the configuration-correlation diagram brings us closer to the goal of predicting how the total electronic energy varies along the reaction coordinate. However, we still must consider the fact that configurations of the same symmetry must be combined (in the configuration interaction step) to give the correct electronic wave functions. In quantitative calculations done on modern computers, the Slater-Condon rules (see Condon and Shortley, 1957, or Cook, 1978) are used to evaluate the Hamiltonian matrix elements

$$H_{ij} = \langle \Phi_i | H | \Phi_j \rangle \tag{3.11}$$

between the important configurations $\{\Phi_i\}$ whose overall space and spin symmetry is correct. The eigenvalues of the H matrix then give the *total* electronic energies of those states that arise from the configurations $\{\Phi_i\}$. These total state energies, when plotted as functions of the reaction coordinate, generate the state-correlation diagram (SCD), which finally allows something to be said about the shape of the potential energy surfaces along the reaction coordinate—in particular, whether large or small reaction barriers are expected.

If *ab initio* calculations are not being done on a computer, a qualitatively correct picture of the state-correlation diagram can still be achieved by using the configuration-correlation diagram. The reasoning is that, when the energies of two configurations *cross* on the configuration-correlation diagram, the states that arise from the mixing of these two configurations will have energies that aviod one another because of configuration interaction (see Shavitt, 1977). Thus, simply by converting all of the crossings that occur in the configuration-correlation diagram to avoided crossings, an approximate state-correlation diagram is obtained.

Before considering how a state-correlation diagram for a chemical reaction is used, it is valuable to review the essential characteristics of the reaction coordinate. It is a totally symmetric motion on the potential surface, except when ψ_0 is degenerate or when low-lying excited states of another symmetry are present that can couple ($\langle \psi_k | \partial V / \partial Q | \psi_0 \rangle$) to ψ_0, in which cases the reaction coordinate becomes symmetric once the symmetry is lowered. The important point is that, by labeling the wavefunctions with *only* those symmetry elements that are preserved along the entire reaction path, the reaction coordinate is *always* symmetric and, hence, the symmetry of ψ_0 remains constant. This means that whenever we guess a reaction coordinate, the symmetries of the orbitals, configurations, and states should be labeled using only those symmetry elements present at all points on the reaction path. For example, in considering the C_{2v} insertion of an atom (say Mg) into the bond of H_2, only the elements of the C_{2v} point group must be used, which means that the ^1S, $^3P_{2,1,0}$, and ^1P states of Mg must be labeled according to how they transform under C_{2v}. Being able to do this is crucial to the use of symmetry correlation concepts as a tool for understanding reactivity.

3.4. Approximate Symmetry

In this section one more point concerning the preserved symmetry elements will be made. The symmetries of the active orbitals (those orbitals involved in the bond-breaking and bond-forming process) are determined by the potential energy field influencing the electrons in these orbitals. This field depends in turn upon how the nuclei and the passive-occupied orbitals are arranged in space. However, those nuclei and passive orbitals that are spatially far from an active orbital will have little influence on the potential field at this active site. As a result, the shape (nodal characteristics and symmetry) of this active orbital will be little influenced by nuclei and orbitals that are far from it. For example, we do not expect the carbonyl π and π^* orbitals of H_2CO (formaldehyde) to be qualitatively different from those of $(H_3C)_2CO$ (acetone) or even $H_3C(CO)H$ (acetaldehyde). In fact, we expect the π and π^* orbitals to maintain their odd character under reflection through the plane containing the C(CO)H group to a very high extent. Certainly the *quantitative* nature of the π orbital, which is more highly localized on the oxygen, and the π^* orbital, which is polarized toward the carbon, will be differently influenced by substituents. However, the basic orbital *nodal characteristics*, which is really the most important aspect of symmetry used, remains largely intact. Thus, approximate *local* symmetry is almost as good as true overall molecular symmetry.

Part 2
Applications to Thermal Reactions

In Chapters 1–3 the theoretical foundations were laid that are needed to understand the energetics of reactions taking place on a single potential energy surface. It would be quite difficult to apply those concepts in a rigorous quantitative fashion to any reaction involving a potential energy surface having more than a few degrees of freedom. The process of walking along the reaction coordinate requires that the potential surface be explored and that the precise nature of the reaction coordinate Q_r be determined. Although such calculations have recently become feasible for systems containing three or four atoms, one almost never knows *exactly* how to walk along Q_r for reactions of more complicated molecules. Therefore, in most applications of symmetry-conservation concepts, a reaction path is *postulated*, and one attempts to explore how the orbital, configuration, and state energies vary along this path. One hopes that, by choosing a path that gives rise to favorable overlap of the important orbitals of the reactant species, the postulated path is close to the true reaction coordinate.

Another problem is that, in addition to being able to know the reaction coordinate precisely, reaction pathways other than those chosen may be available. For example, if the one-step four-center reaction of H_2 with I_2 to produce 2HI were examined, a reasonable conclusion would be that the reaction has a high activation energy. However, this conclusion does not eliminate the possibility that HI can be formed by some other mechanism, so it is important to explore all pathways that might yield the desired products. The symmetry methods illustrated in this chapter and those for photochemical reactions in Chapter 7 can be used to analyze any single reaction step that involves breaking old bonds and *simultaneously* forming new product bonds. If a proposed reaction path causes reactants to give rise to products in a single step, the reaction is termed *concerted*. The symmetry rules can be applied to the single step of such concerted reactions to predict directly whether the reaction would have a large activation energy. For stepwise reaction mechanisms, the symmetry rules must be applied to each step. If any step in the reaction is predicted to have a large activation energy, the overall reaction would not be expected to proceed with great speed.

It is suggested that readers who have not recently made use of point-group symmetry tools such as character tables, direct products, and projection operators read Appendix C before beginning the examples treated in Chapter 4.

Chapter 4

Examples for Analyzing Ground-State Thermal Processes

In this chapter several examples of applications of the symmetry arguments are presented.

4.1. Simple Predictions from Orbital-, Configuration-, and State-Correlation Diagrams

The thermal addition reaction of a nitrogen molecule and a hydrogen molecule to yield *cis*-diimide is a straightforward problem.

We begin by considering only the orbitals directly involved in changing the bonds and assume that C_{2v} symmetry is preserved during the reaction path—this is the guess of the reaction coordinate. The steps for attacking the problem are the following:

1. An orbital-correlation diagram is drawn for the proposed reaction path.
2. A hypothesis is made for the likely (energetically favorable) configurations, and a configuration-correlation diagram is constructed.
3. The state-correlation diagram is made and is then used to determine whether the reaction is thermally allowed along this reaction path.

If we *assume* that the reaction coordinate involves the C_{2v} approach as shown in Figure 4-1, the active orbitals are the bonding and antibonding σ_{HH} and π_{NN} orbitals. Notice that we are not trying to deduce the reaction coordinate but merely proposing a reaction coordinate and observing whether a high barrier to *that* reaction is expected. The basic approach is to try all reaction pathways that we believe to be likely for good bond formation. If we then assume that the potential energy surface can be smoothly interpolated between

36   CHAPTER 4

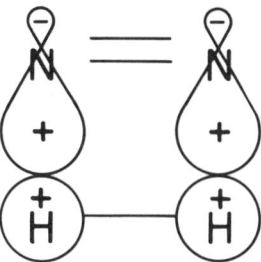

Figure 4-1
Cis addition path.

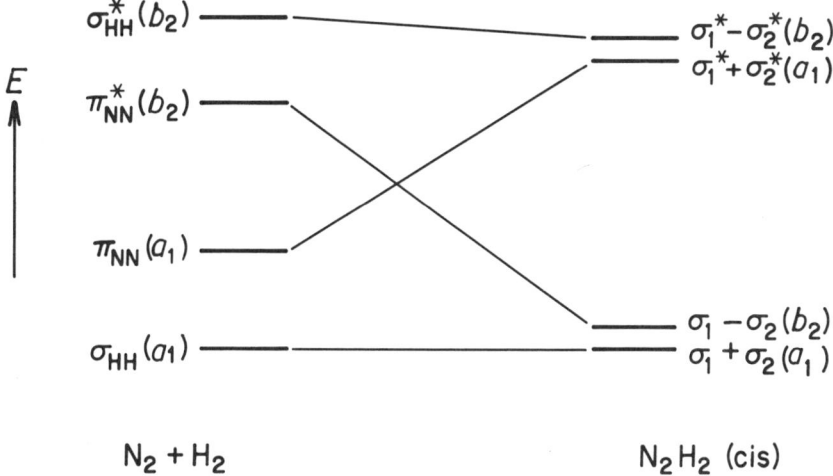

Figure 4-2
Orbital-correlation diagram for cis addition.

these postulated pathways, which have some symmetry, then a qualitative picture of the full surface can be obtained.

Since nitrogen is more electronegative than hydrogen, the expected energy ordering of the N_2 and H_2 orbitals is that shown in Figure 4-2. The spacing between the bonding and antibonding H_2 orbitals is larger than for the N_2 π bond energy. In more complicated situations, one often resorts to using information about valence ionization potentials to provide, via Koopmans' theorem (Pilar, 1961), the ordering of the orbital energies for reactants and products. Figure 4-2 also yields the symmetries of the orbitals of both $N_2 + H_2$ and cis-HNNH. Note that the *orbital*-correlation diagram involves a crossing among the second and third molecular orbitals. This fact alone does not mean that the

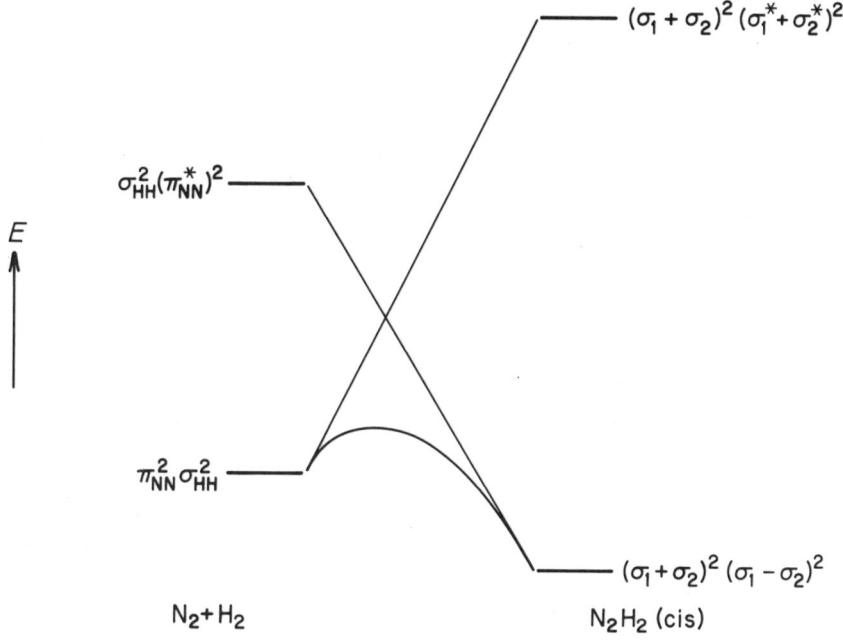

Figure 4-3
Configuration-correlation diagram and state-correlation diagram for cis addition.

reaction $N_2 + H_2 \rightarrow$ HNNH is forbidden (remember that only cis-HNNH is being treated here because we assume C_{2v} symmetry); we still have to look at the configuration-correlation diagram and the state-correlation diagram.

Since the ground electronic states of N_2 and H_2 possess double occupancy of the bonding π_{NN} orbital and the bonding σ_{HH} orbital, the $\pi_{NN}^2 \sigma_{HH}^2$ configuration, which has $(a_1^2 a_1^2) = {}^1A_1$ symmetry, should be important. At the other end of the assumed reaction coordinate is cis-HNNH, which should be dominated by the $(\sigma_{NH_1}^2)(\sigma_{NH_2}^2) = (a_1^2)(b_2^2) = {}^1A_1$ configuration. The fact that these two *configurations* do not correlate is shown in Figure 4-3, which also shows the avoided configuration crossing that gives rise to the state-correlation diagram. Notice that configurations are correlated according to the symmetries of the orbitals that are occupied in the configurations rather than by their overall space-spin symmetry. The steepness of the two configuration energy lines shown in Figure 4-3 is determined by the relative energies of the two dominant configurations at the extremes of the reaction coordinate. For example, the $\pi_{NH}^2 \sigma_{HH}^2$ configuration, which correlates to $\sigma_{NH}^2 \sigma_{NH}^{*2}$, is expected to be very high in energy (since both NH bonds are broken) on the diimide side of the reaction. Based upon the above state-correlation diagram, a substantial barrier to this thermal reaction is expected.

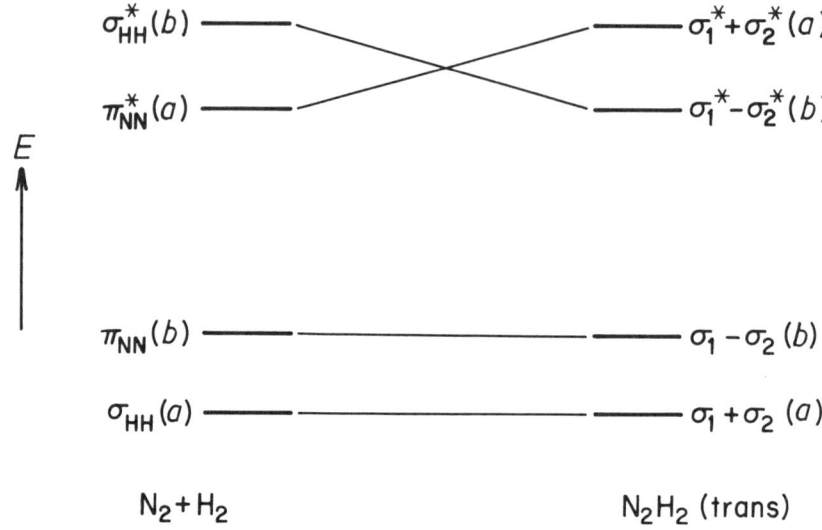

Figure 4-4
Orbital-correlation diagram for trans addition.

We must also consider whether the interaction between the two configurations that cross would be large enough to produce such strongly avoided crossings that the state-correlation diagram would show no barrier. This is not likely because the configuration coupling is caused only by the r_{ij}^{-1} terms in h_e (since the configurations differ by *two* orbitals as discussed in Appendix A), and these electronic interaction terms are usually quite small. Hence, configuration pairs that differ by two orbitals relative to one another should display *weakly* avoided crossings. Notice that the orbital crossing (noncorrelation) leads to the configuration noncorrelation, from which a high barrier is predicted. The above analysis shows that this reaction is thermally forbidden. The word *thermal* is employed because the molecular orbitals are occupied in a way that is appropriate to the ground states of reactants and products, and the system is considered to move on this ground-state surface. The reaction is forbidden only because a large symmetry-imposed activation barrier to this thermal reaction should be present.

We now consider the reaction $N_2 + H_2 \rightarrow$ HNNH(trans). The relevant point group is now C_2 (where the C_2 rotation axis is perpendicular to the plane of the molecule, and the orbital correlation diagram is given in Figure 4-4. Now the orbital symmetries correlate differently, so that the configuration-correlation diagram (Figure 4-5) does not involve configuration crossing (i.e.,

EXAMPLES FOR GROUND-STATE THERMAL PROCESSES

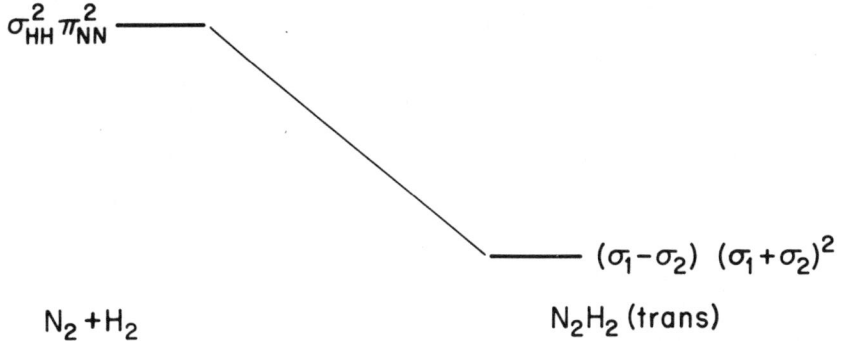

Figure 4-5
Configuration-correlation diagram and state-correlation diagram for trans addition.

$\pi_{NN}^2 \sigma_{HH}^2$ correlates directly with $\sigma_{NH_1}^2 \sigma_{NH_2}^2$). Hence, the formation of trans-HNNH via a C_2 reaction coordinate should involve no significant symmetry-imposed barrier. Thus, N_2H_4 should thermally fragment via a *trans* reaction coordinate, and H_2 should attack N_2 in a trans manner. This latter prediction is based upon symmetry considerations alone, and it does *not* mean that such a trans attack would occur easily—for example, in a collision. In fact, such a collision is not likely to be successful, since the bond length of the H_2 molecule is so short that a great deal of energy would be required just to stretch the H_2 sufficiently to make formation of the two new NH bonds feasible.

The above symmetry considerations include nothing about geometrical factors or the overall reaction thermodynamics (such as bond strengths of the reactant and product); symmetry only makes requirements on the nodal patterns of the important orbitals. It is essential to keep this point in mind in all subsequent problems.

Notice also that, because the C_{2v} point group has no degenerate representations, nowhere (except at infinite separation) along the hypothetical reaction path do any orbitals or states become degenerate. Hence, questions involving first-order Jahn-Teller instability do not occur in this example.

Only two possible reaction paths have been considered, and it is reasonable to wonder whether collisions that are slightly non-C_{2v} can yield cis-diimide, because in the C_s point group, which would rigorously pertain to such a collision, the a_1 and b_2 orbitals both have a' symmetry. As a result, the rigorous orbital-correlation diagram would *not* include an orbital crossing and, hence, the configuration-correlation diagram and state-correlation diagram would not predict a forbidden reaction. However, this rigorous analysis is incorrect! The concept of near symmetry also allows us to apply conclusions based upon

C_{2v} symmetry analysis to collisions that involve near-C_{2v} geometries. That is, *near* the C_{2v} geometry that was analyzed, a symmetry-imposed barrier, whose precise shape and height will vary somewhat as one moves further from C_{2v} geometry, will still be present. As will be seen in later examples, it is not the rigorous symmetry that actually gives rise to orbital symmetry nonconservation but the nodal characteristics of the orbitals. These nodal patterns remain even when rigorous symmetry is lost.

4.2. An Example of Unimolecular Decomposition

This example makes use of more quantitative data on ionization energies to order the molecular orbitals. If such data is available, it is certainly wise to use it, because then one can make more quantitative predictions about the thermal and photochemical behavior of a system. As an example, the thermal decomposition of formaldehyde to carbon monoxide and a hydrogen molecule

$$H_2CO \rightarrow H_2 + CO$$

is analyzed. A minimum-basis molecular-orbital calculation has been carried out (Cook, 1978; Pilar, 1968; Schaefer, 1971) and has yielded the orbital energies (in eV) given in Table 4-1. We begin by postulating a reaction coordinate, by assuming that C_{2v} symmetry is preserved during the decomposition, and proceed through the following steps:

1. The symmetry of the molecular orbitals of the formaldehyde molecule, the hydrogen molecule, and carbon monoxide are classified according to the point-group symmetry that is preserved during the reaction path.
2. An orbital-correlation diagram for this reaction path is drawn.
3. A configuration-correlation diagram is constructed, and the state-correlation diagram is used to determine whether the thermal decomposition of formaldehyde is allowed.

The irreducible representations of the formaldehyde molecular orbitals given in Table 4-1 determine the coordinate system to be used. The $1a_1$ and $2a_1$ molecular orbitals are the $1s$ orbitals on oxygen and carbon, respectively. All a_1 orbitals have the symmetry of the z coordinate axis. The $1b_1$ orbital is unique among the occupied molecular orbitals and, hence, must represent the π molecular orbitals of the carbonyl group that are aligned along the y axis, lying perpendicular to the plane of the formaldehyde molecule (Figure 4-6). The symmetry labels that can thus be attached to the active molecular orbitals are the following:

The hydrogen molecule. $1\sigma_g(a_1)$, $1\sigma_u(b_2)$

TABLE 4-1

Orbital Energies in eV For Formaldehyde, Carbon Monoxide, and Hydrogen

Formaldehyde (C_{2v})		Carbon monoxide ($C_{\infty v}$)		Hydrogen ($D_{\infty h}$)	
$1a_1$	−553.4	1σ	−553.3	$1\sigma_g$	−17.51
$2a_1$	−302.8	2σ	−302.7	$1\sigma_u$	20.05
$3a_1$	−36.52	3σ	−35.82		
$4a_1$	−23.02	4σ	−19.92		
$1b_2$	−18.48	1π	−15.78		
$5a_1$	−15.09	5σ	−13.09		
$1b_1$	−12.22	2π	7.07		
$2b_2$	−10.28	6σ	25.36		
$6a_1$	3.99				
$2b_1$	7.10				
$7a_1$	19.75				
$3b_2$	23.07				

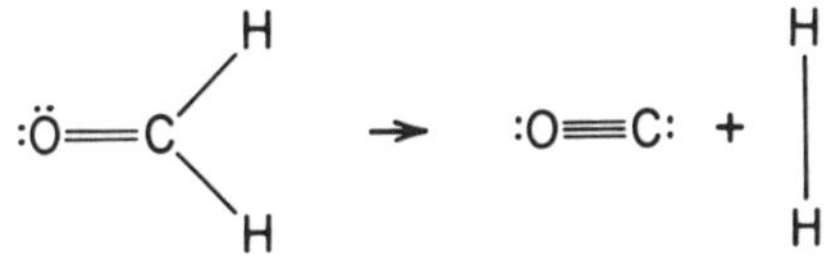

Figure 4-6
Decomposition of formaldehyde.

Carbon monoxide. All of the σ and lone-pair orbitals transform as a_1; the π orbitals transform as b_1 and b_2; the π^* orbitals also transform as b_1 and b_2.

Formaldehyde. The CO σ and one of the CH σ bonds transform as a_1, the other CH σ bond as b_2, the nonbonding orbitals on oxygen as a_1 and b_2, and the CO π and π^* orbitals as b_1.

The orbital-correlation diagram that results from simply connecting the orbital lists of H_2CO and $H_2 + CO$ is given in Figure 4-7. Notice that several orbital crossings are present, so the possibility exists that an activation barrier will be predicted. By using the 16 electrons to occupy the lowest 8 molecular orbitals of H_2CO (since we are considering the thermal or ground-state reaction)

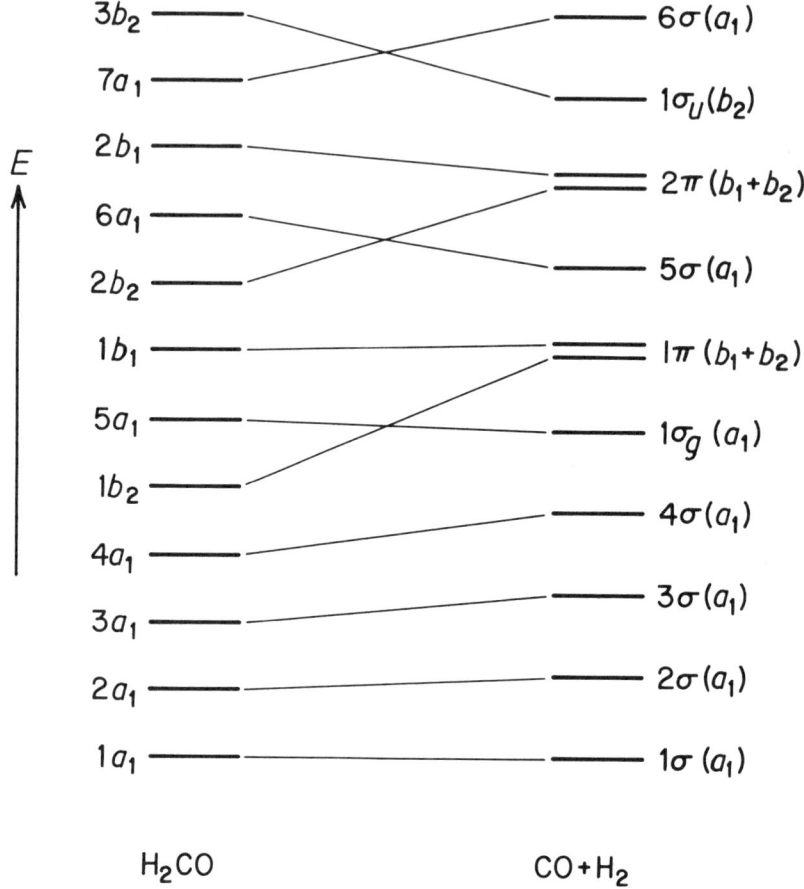

Figure 4-7
Orbital-correlation diagram for decomposition of formaldehyde.

the $2b_2^2$ part of the resulting configuration correlates with a $2\pi^2$ occupancy of CO (i.e., double occupancy of an antibonding π_{CO}^* orbital). The 2π orbital of the ground state of CO is not doubly occupied. The crossing of the $5a_1$ and $1b_2$ orbital energies is not relevant, for both of these orbitals are doubly occupied in the ground states of both the reactant and the product.

The configuration-correlation diagram shown in Figure 4-8 results when the lowest 8 orbitals of $CO + H_2$ and of H_2CO are occupied by the 16 electrons. The two configurations that cross do have the same overall symmetry ($^1A_1$), so they can mix to give the state-correlation diagram also shown in Figure 4-8 by the curves undergoing avoided crossing. However, the existence of the orbital and configuration crossings should give rise to a substantial barrier to this

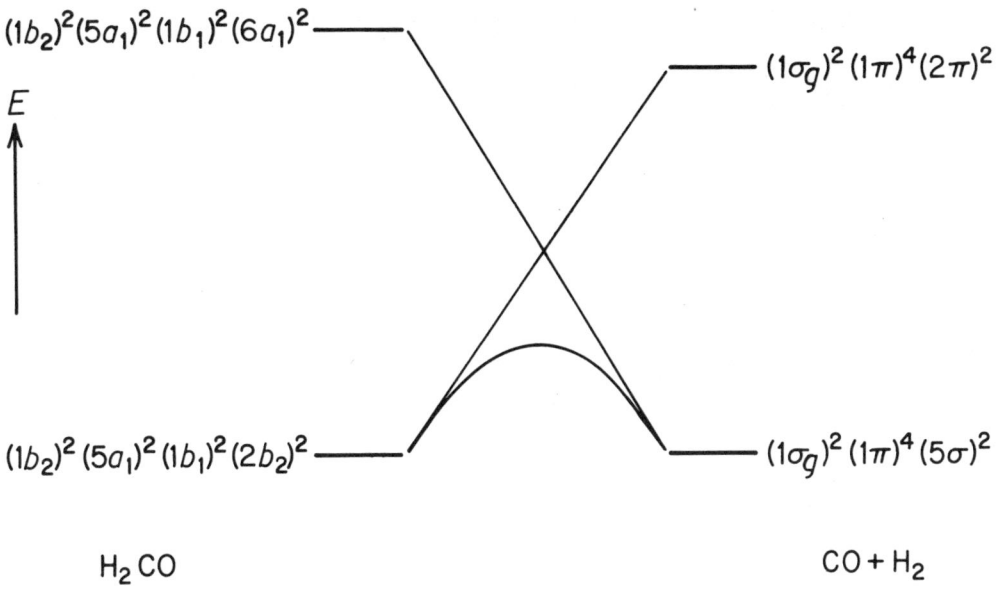

Figure 4-8
Configuration-correlation diagram and state-correlation diagram for decomposition of formaldehyde.

thermal reaction. Since the two configurations that cross differ with respect to one another by two electron occupancies, only the electron-electron interaction terms (r_{ij}^{-1}) in h_e couple them together. Hence, the barrier predicted by the configuration crossing should not be substantially lowered in going to the state-correlation diagram.

Because of the low point-group symmetry (C_{2v}), degenerate representations are not possible; therefore, first-order Jahn-Teller effects do not come into play. Notice also that if the reaction were to occur along a path that preserves only the one plane of symmetry (C_s point group), then both b_2 and a_1 orbital become a'. Consequently, the reaction becomes orbitally allowed (since now the orbitals of reactants and products correlate) and, hence, allowed in the configuration- and state-correlation diagrams. Thus, the high barrier predicted for the C_{2v} path will probably be reduced when moving away from the C_{2v} path. However, the barrier should not *suddenly* disappear when moving slightly away from the C_{2v} symmetry because approximate symmetry is still present. In contrast, if we imagine distortions of H_2CO that make the molecule nonplanar but preserve the plane of symmetry bisecting the HCH bond angle, the crossing of the $2b_2$ and $6a_1$ orbitals would still occur, since these two orbitals have different symmetry under this plane. Hence, this kind of molecular deformation should not lower the symmetry-imposed barrier substantially.

4.3. A Degenerate Case with Jahn-Teller Effects

In this section a case in which degenerate orbitals can occur is examined. We ask whether the π orbital structure of the cyclopropenyl radical (C_3H_3) favors an equilateral triangular structure. The $(C_3H_3)^+$ or $(C_3H_3)^-$ forms will also be considered. At the equilateral-triangle geometry, the p_π orbitals on the three carbon atoms can be combined to yield one orbital having a_1 symmetry (C_{3v} point group)

$$a_1 = P_A + P_B + P_C$$

and two degenerate orbitals of e symmetry

$$e = \begin{bmatrix} P_A - P_C \\ 2P_B - P_A - P_C \end{bmatrix}$$

These symmetry-adapted orbitals can be generated by applying the projectors (see Appendix C or Cotton, 1963)

$$\mathbf{P}_i = \frac{1}{g} \sum_R \chi_i(R) R$$

to the individual atomic basis orbitals (P_A, P_B, P_C), in which $\chi_i(R)$ is the character of irreducible representation i. The two degenerate e orbitals that would be obtained are not orthogonal. After orthogonalization the two e orbitals given above are obtained. The a_1 orbital is bonding between all three carbon atoms, whereas the e orbitals are antibonding. Because only the effects of the π orbitals are examined here, all of the factors that determine the shape of C_3H_3 are not considered. If the σ-bonding effects are sufficiently strong (i.e., if the σ force constants are large enough), the π-orbital effects treated here will be negligible. However, they are still interesting and instructive to study.

What predictions can be made within these limitations? First, $(C_3H_3)^+$ should be stable at the equilateral-triangle geometry, since it would have both of its π electrons in the bonding a_1 orbital; thus, ψ_0 would be nondegenerate. Second, C_3H_3 should be first-order Jahn-Teller unstable, since the configuration $(a_1)^2 e$ is degenerate. The kind of vibration that will distort the equilateral-triangle geometry is predicted (by forming the direct product of $\psi_0^* \psi_0$ (Cotton, 1963) from the first-order Jahn-Teller matrix element $\langle \psi_0 | \partial h_e / \partial Q | \psi_0 \rangle$ to have ($e \times e = a_1 + e + a_2$) or e symmetry (for this point group, a_2 is a rotation and a_1 a symmetric ring distortion that would not break the degeneracy of ψ_0) that corresponds to distorting to C_{2v} symmetry by pulling one of the CH groups away from the other two. Consequently, equilateral triangle C_3H_3 is unstable and cannot even be a true transition state (since it has a nonzero value of slope). Third, $(C_3H_3)^-$ should not be stable at the equilateral-triangle geometry for two possible reasons: (1) it has two electrons in the antibonding e orbitals that would "cancel" any π bonding due to the two a_1 electrons (but the σ bonds would remain intact). (2) The configuration $(e)^2$ might be Jahn-Teller unstable. The $(e)^2$ configuration has symmetry components $^1A_1 + {}^3A_2 + {}^1E$, of which, by Hund's rules, the $^3A_2$ would be the lowest energy state. This triplet state is not spatially degenerate and is therefore not first-order Jahn-Teller unstable.

The above symmetry- and spin-term symbols for $C_3H_3^-$ are obtained by forming, for the singlet states (which have antisymmetric two-electron spin functions), the symmetric direct product $(e \times e)_+$; for the triplet (which has an even spin function), one forms the antisymmetric direct product $(e \times e)_-$. The characters $(\chi_\pm)$ for these two kinds of direct products are given in terms of the characters (χ) of the e representation of the individual orbitals appearing in the e^2 configuration as

$$\chi_\pm(R) = \frac{1}{2}[\chi^2(R) \pm \chi(R^2)].$$

Within the C_{3v} point group,

	E	$2C_3$	$3\sigma_v$
χ_+	3	0	1
χ_-	1	1	-1

Then, the $\chi_\pm$ are decomposed into their individual representations using the projections (Cotton, 1963)

$$n_i = \frac{1}{g}\sum_R \chi_i(R)\chi_\pm(R)$$

(g is the group order 6, and the χ_i are the characters of irreducible representations), which indicate the number of times (n_i) representation i occurs in $\chi_\pm$. It is important to use these symmetric and antisymmetric direct products when dealing with two *equivalent* electrons. If the electrons were in different sets of degenerate orbitals ($1e$, $2e$), then they are nonequivalent and you can use the usual direct product (Cotton, 1963).

The lowest energy state of $(e)^2({}^3A_2)$ is nondegenerate; thus, $(C_3H_3)^-$ is *not* unstable by a first-order Jahn-Teller distortion. Its unstable nature is an effect of the antibonding nature of the e orbitals. *If* the excited states (^1E, $^1A_1$)

did not have different spin symmetry from the $^3A_2$ ground state, then quadratic terms in the Jahn-Teller theory should come into play through the $\langle\psi_0|\partial h_e/\partial Q|\psi_k\rangle$ factors. The distortion that should contribute would have either $A_2 \times A_1 = A_2$ or $A_2 \times E = E$ symmetry. The C_{3v}-character table shows that the molecule has no A_2 vibration, though it does have an E vibration. Hence, an E distortion should occur. Of course, this analysis is nothing but an exercise in futility because the different spin symmetries of the $^3A_2$ and ($^1A_1$, ^1E) states would make the $\langle\psi_0|\partial h_e/\partial Q|\psi_k\rangle$ integrals vanish—unless spin orbit effects were very large—and this is not likely for $(C_3H_3)^-$.

4.4. The Bond-Symmetry Rule—Another Jahn-Teller Case

In this section we consider whether two ethylene molecules will combine to give cyclobutane if they collide in a D_{2h}-symmetry manner (in a head-on fashion such that their π orbitals bump into one another directly) and then examine the analogous exchange reaction $H_2 + D_2 \rightarrow 2HD$.

The orbital-correlation diagram for both of these reactions is shown in Figure 4-9, in which each orbital is labeled according to its D_{2h} symmetry (and according to the the D_{4h} symmetry that appears once the four hydrogen atoms are equivalently located in the latter reaction). Notice, for $H_2 + D_2$, that the b_{2u} and b_{3u} orbitals cross at D_{4h} symmetry where they are symmetry-degenerate (e_u). These orbitals also cross in the case of ethylene, but not at D_{4h} symmetry. As a result of these orbital crossings, the ethylene $\pi^2\pi^2(a_g^2b_{2u}^2)$ and $\sigma^2\sigma^2(a_g^2b_{3u}^2)$ configurations also cross one another—they do not correlate (Figure 4-10). The corresponding configurations also cross in the case of H_2 and D_2. Since both of these configurations have ^1A_g symmetry, the state-correlation diagram will show an avoided crossing once they mix to give rise to two ^1A_g state wave functions. A substantial barrier (an avoided crossing) to both of these reactions should be present, but it should be weakly avoided because the two participating configurations differ by two orbitals, so only the r_{ij}^{-1} can couple them.

If the D_{4h} geometry were proposed as a possible transition state for the hydrogen-exchange reaction, several kinds of motion that might be important in distorting this geometry must be considered. At D_{4h}, the configuration $a_{1g}^2e_u^2$—through the symmetric direct product for this singlet reaction—gives rise to $^1A_{1g} + ^1B_{1g} + ^1B_{2g}$ electronic states. Quantum chemical calculations indicated that the $^1B_{1g}$ state is the lowest singlet state. Thus, this $D_{4h}{}^1B_{1g}$ state is not first-order Jahn-Teller unstable, and we cannot conclude that the slope of the surface is nonzero along any nonsymmetric direction. Hence, it is possible that the square geometry is a transition state. Furthermore, quadratic Jahn-Teller effects involving $\langle ^1B_{1g}|\partial h_e/\partial Q|^1A_{1g}\rangle$ could give rise to a $B_{1g} \times A_{1g} = B_{1g}$ distortion of the molecule.

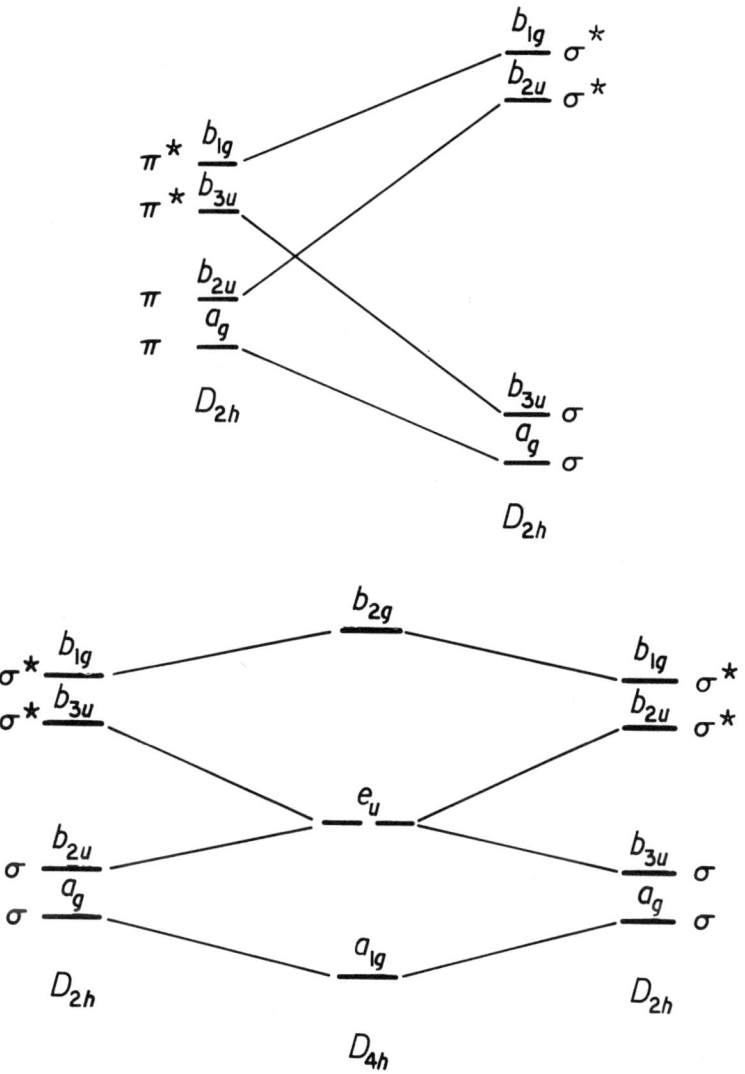

Figure 4-9
Orbital-correlation diagrams for (a) dimerization of ethylene and (b) H_2–D_2 exchange.

The other distortion $B_{1g} \times B_{2g} = A_{2g}$ cannot occur because the D_{4h} point group has no A_{2g} vibration (Cotton, 1963). Thus a $D_{4h}H_4$ transition state might be unstable with respect to a pseudo-Jahn-Teller effect involving a B_{1g}

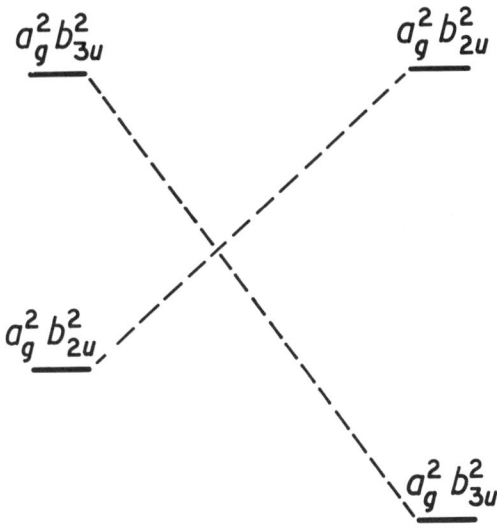

Figure 4-10
Configuration-correlation diagram for dimerization of ethylene.

vibration. The term *pseudo*-Jahn-Teller is used rather than second-order Jahn-Teller, because the excited state ($\psi_k = {}^1A_{1g}$) arises by electron rearrangement between the two degenerate e_u orbitals (analogous to the atomic orbital configuration p^2 giving rise to ${}^3P$, ${}^1D$, and ${}^1S$ states) and not from electron promotion into an excited orbital. One cannot conclude that the square geometry is definitely pseudo-Jahn-Teller unstable because, from what has been said, it is not clear whether the negative-curvature terms caused by coupling of the ${}^1B_{1g}$ and ${}^1A_{1g}$ states are larger in magnitude than the positive-curvature terms described in equation 2.20—more quantitative information about these coupling matrix elements is needed to make this prediction.

This example—two ethylenes combining to give cyclobutane—illustrates the concept called the *bond-symmetry rule* (Pearson, 1976). This rule states that a reaction will be orbitally forbidden—that is, involve orbital crossing problems—unless the symmetries of the bonds plus lone pairs broken in the reactants match the symmetries of bonds plus lone pairs formed in the products. For example, the rearrangement of benzene—which has one occupied a_1b_1 and $b_2\pi$ orbital (see Cotton, 1963, and section 7.5)—to give dewarbenzene (which has two occupied a_1 orbitals and one occupied b_2 orbital) is forbidden. Also the C_{2v} decomposition of

$$\begin{array}{c} H \\ \diagdown \\ C = O \\ \diagup \\ H \end{array}$$

to give

$$\begin{array}{c} H \\ | \\ H \end{array} + :C = O$$

is forbidden, since the reactant has two a_1 and two b_2 orbitals (CH bonds and oxygen lone pairs), whereas the products have one b_2 orbital (the CO π bond) and three a_1 orbitals (HH and the :CO: lone pairs).

The addition of a halogen to the 1 position of dewarbenzene

reduces the symmetry from C_{2v} to C_s. As a result, the reactant has two a' and one a'' orbitals, as does the halobenzene. Thus, the reaction becomes allowed, though the reaction rate is still very low (Pearson, 1976, p. 90). The reason is that the halogen does not have a strong influence on the active orbitals, so C_{2v} symmetry (in which the reaction is forbidden) is approximately valid.

The bond-symmetry rule is, or course, nothing but a short cut for recognizing when nonconservation of occupied-orbital symmetry will occur. Whenever the symmetries of the occupied orbitals of reactants and products do not match, the orbital-correlation diagram will display a crossing of occupied and unoccupied orbitals, which then causes nonconservation of symmetries.

Use of this (or any other) symmetry rule requires that only symmetry elements that are preserved along the full reaction path are used to label the orbitals. Only for these symmetry elements is the reaction coordinate Q_r symmetric, and this is necessary if the orbital, configuration, and state symmetries are to remain constant (i.e., be correlated by symmetry) along the reaction path. For example, in considering the opening of

to give

two possible "reaction paths" are commonly examined (Woodward and Hoffman, 1970). Two kinds of ring opening—*conrotatary* and *disrotatary*—must be considered. In conrotatory opening, both CH_2 groups are twisted in the same direction (e.g., clockwise). When this motion is used as a reaction coordinate, the only symmetry element preserved throughout the reaction is a C_2 axis

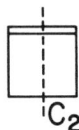

The occupied σ and π orbitals of the reactant and the two occupied π orbitals in the 1,3-butadiene product are even and odd, respectively, with respect to this C_2 axis. Hence, the bond-symmetry rule indicates that conrotatory ring opening is orbitally allowed.

In disrotatory opening, the two CH_2 groups move in opposite directions. Such motion preserves a reflection plane σ_v that runs down the middle of the molecule

Under this symmetry element the σ and π orbitals of the reactant are both even, whereas the π orbitals of the product are even and odd. Hence, disrotatory ring opening is orbitally forbidden.

The two different reaction paths (conrotatory and disrotatory) preserve different symmetry elements and therefore lead to different symmetry predictions.

4.5. Breaking of Single Homonuclear and Heteronuclear Bonds

Before considering more sophisticated examples, we return for a moment to the most elementary reactions—those in which only a single bond is fragmented, as in the dissociation of H_2 and HCl. These simple bond-breaking reactions serve as an excellent test of our concepts, for if these ideas are valid, they must certainly apply for these reactions. In these cases, only one active bond is broken, so the only orbitals to be considered in constructing an orbital-correlation diagram are the bonding and antibonding orbital pairs (σ_g and σ_u for H$_2$, σ and σ^* for HCl). These pairs of orbitals never cross, so such bond breaking cannot be orbitally forbidden. It is important to note that not being forbidden does not mean that the bond breaking costs no energy but instead that symmetry constraints do not impose an *additional* energy barrier to such reactions beyond simple thermochemical energy requirements. In general, the meaning of the symmetry rules forbidding or not forbidding a reaction is only the presence or absence, respectively, of an *additional* energy barrier. The overall thermodynamic stability of reactants and products, which has nothing to do with symmetry, must always be included when one attempts to guess *total* reaction activation barriers.

So far, we have concluded that for breaking (or forming) simple bonds, an orbital-symmetry-related activation barrier will not exist, because the bonding and antibonding orbitals do not cross. We now consider the configuration-correlation diagram for such a reaction, treating the homonuclear and heteronuclear molecules separately. For H$_2$, the available orbitals (σ_g and σ_u) can be occupied in several different ways—σ_g^2 and σ_u^2 give $^1\Sigma_g$ configurations, and $\sigma_g\sigma_u$ can give $^{1,3}\Sigma_u$. Since the ground state of H$_2$ has $^1\Sigma_g$ symmetry, only the σ_g^2 and σ_u^2 configurations can play a role. The energies of these two configurations differ greatly for internuclear distances (R) near equilibrium, but they become degenerate when $R \to \infty$. Therefore, a strong configuration interaction should exist between σ_g^2 and σ_u^2 for large values of R.

We now analyze the behavior of the σ_g^2 and σ_u^2 configurations at large R to see why they become degenerate. Using the facts that the unnormalized molecular orbitals can be expressed in terms of the 1S atomic orbitals as

$$\sigma_g \cong 1S_A + 1S_B$$

$$\sigma_u \cong 1S_A - 1S_B$$

for large R, the two relevant Slater-determinant wavefunctions can be written as (Pilar, 1968)

$$|\sigma_g^2| = \sigma_g(1)\sigma_g(2)(\alpha\beta - \beta\alpha)2^{-1/2}$$
$$= [1S_A1S_A + 1S_B1S_B + 1S_A1S_B + 1S_B1S_A](\alpha\beta - \beta\alpha)2^{-1/2}$$

and

$$|\sigma_u^2| = [1S_A1S_A + 1S_B1S_B - 1S_A1S_B - 1S_B1S_A](\alpha\beta - \beta\alpha)2^{-1/2}.$$

Notice that, at large R, both of these configurations contain equal mixtures of ionic (e.g., $1S_A1S_A$) and covalent (e.g., $1S_A1S_B$) terms. Neither configuration properly describes 2H at $R \to \infty$; they both attempt to describe $\frac{1}{2}[2H + H^+, H^-]$, whose energy lies above that of 2H by one-half the ionization potential of H minus the electron affinity of H. However, at large R the configuration-interaction function $(1/\sqrt{2})[|\sigma_g^2| - |\sigma_u^2|]$ would contain only covalent (2H) terms and its orthogonal partner $(1/\sqrt{2})[|\sigma_g^2| + |\sigma_u^2|]$ would describe (H$^-$, H$^+$). Recall that $\sigma_g \sigma_u$ could not contribute because it has the wrong symmetry. Thus, the breaking of a single *homonuclear* bond has no symmetry-caused activation energy and requires mixing of σ^2 and $(\sigma^*)^2$ configurations that are doubly excited relative to one another and that mix very strongly because the σ and σ^* orbitals are degenerate for large R.

For *heteronuclear* bonds the situation is somewhat different. For example, in HCl the bonding (σ) and antibonding (σ^*) orbitals do not cross, and at large R they do not become degenerate. However, the σ^2 configuration and the $\sigma\sigma^*$ (singlet) configurations do cross as R varies. Near the equilibrium value of R, σ^2 represents the σ bond of HCl, whereas $\sigma\sigma^*$ describes a dissociative excited state. At large values of R the σ orbital becomes $3p$ Cl and σ^* becomes $1s$ H, so σ^2 describes (H$^+$, Cl$^-$) and $\sigma\sigma^*$ represents (Cl, H). Therefore, a configuration interaction should be important in describing breakage of the HCl bond. Notice that, in contrast with the homonuclear case, *singly excited* configurations play the dominant role. As a result, the operator $\partial h_e / \partial Q$ can effectively couple the σ^2 and $\sigma\sigma^*$ configurations and thereby produce a smooth (barrier-free) transition from HCl to H + Cl as the reaction coordinate varies. In *both* cases, a *strong* configuration interaction (σ_g^2 and σ_u^2 for H$_2$ and σ^2, $\sigma\sigma^*$ for HCl) yields a smooth potential energy curve that displays *no* symmetry-imposed barriers. Symmetry barriers arise only when more than one electron pair undergoes changes in a reaction, since it is only in this more complicated case that orbital crossings (which then produce avoided configuration crossings) can exist along the reaction coordinate.

4.6. The Use of Bonding-Antibonding Orbital Mixing to Predict the Reaction Coordinate

In section 4.4 we showed that, as far as being symmetry-allowed, cyclobutene could open by means of a conrotatory motion to yield 1,3-butadiene. In that example, a reaction path was guessed and then tested for orbital-symmetry conservation. That reaction can also be viewed in a different way. In section 4.5 we showed that breaking bonds utilizes mixing of configurations having antibonding- and bonding-orbital partners. Cyclobutene has, in C_{2v} symmetry, doubly occupied $a_1(\sigma)$ and $b_1(\pi)$ active orbitals and corresponding empty $b_2(\sigma^*)$ and $a_2(\pi^*)$ orbitals. If we proposed to add in configurations (through $\langle \psi_0 | \partial h_e / \partial Q | \psi_k \rangle$) that include one or both of the excitations $a_1 \to b_2$ ($\sigma \to \sigma^*$) and/or $b_1 \to a_2$ ($\pi \to \pi^*$), the symmetry of the distortion coordinate would have to be $a_1 \times b_2 = b_1 \times a_2 = b_2$ because ψ_0 is $^1A_1$ and ψ_k has the symmetry of the direct product of its singly occupied orbitals. On the other hand, $a_1 \rightarrow a_2$ ($\sigma \rightarrow \pi^*$) and $b_1 \rightarrow b_2$ ($\pi \rightarrow \sigma^*$) excitations could be caused by a distortion of a_2 symmetry. The CH_2 twisting motions of cyclobutene, which are needed to break the σ bond between the carbon atoms 1 and 4, have b_1 (disrotatory) and a_2 (conrotatory) symmetry (Cotton, 1963); b_2 motion would be an in-plane deformation of the carbon ring, not a twisting motion. Hence, the conrotatory motion produces proper bonding-to-antibonding orbital mixing and this causes bond breakage. This reaction, in which a new π bond is formed from the termini of a conjugated π system, is called an *electrocyclic reaction*.

Notice that in this example knowledge of the symmetry of the molecular orbitals of the product molecule was not used. Instead, mixing of bonding and antibonding orbitals of the reactants was the approach employed. Consequently, the technique can only predict the kind of motion that can *break* certain bonds and is independent of the bonds that form. This idea of mixing the bonding molecular orbital of one fragment with the antibonding molecular orbital of the other fragment (e.g., $\sigma \rightarrow \pi^*$, $\pi \rightarrow \sigma^*$) is very important. This mixing allows charge density to flow from the old bonds into regions of space (orbitals) that allow these bonds to break while new bonds form. Recall from introductory quantum chemistry (Cook, 1978; Shavitt, 1979; Appendix A) that singly excited configurations are used to describe orbital polarization or orbital relaxation. That is, an excitation of the form $\sigma^2\pi^2 \rightarrow \sigma\pi^2\pi^*$ gives rise to polarization of the σ orbital in a way that mixes in some π^* character. If this polarized orbital ($\sigma + x\pi^*$) looks like a bonding orbital of the product, the reaction is favored. Earlier in this book the *change* in charge density $\Sigma_{k \neq 0} \delta \rho_{k0}$ caused by mixing singly excited configurations was considered, and it was seen that nuclei move to regions of space in which $\Sigma_{k \neq 0} \delta \rho_{k0}$ is positive. Hence, in order to propose that nuclei move from one part of a molecule to another (e.g., when conrotatory and disrotatory openings of cyclobutene were examined), it is necessary to look for single excitations that give positive values of $\Sigma_{k \neq 0} \delta \rho_{k0}$ for those regions of the molecule. If these same single excitations also give rise to $\Sigma_{k \neq 0} \delta \rho_{k0}$ patterns that allow the "new" bonds to form as the "old" bonds are being broken, the *concerted* reaction should be symmetry-allowed. Hence, for cyclobutene, when we mix the $\sigma^2\pi^2$ configuration with the $\sigma\pi^2\pi^*$ and $\sigma^2\pi\sigma^*$, we are trying to allow electron density to flow from σ to π^* and from π to σ^*, respectively. Clearly, these excitations (orbital polarizations) allow the σ and π bonds of the reactants to rupture. Moreover, flow of charge into the π^* orbital produces electron density in an orbital of the form

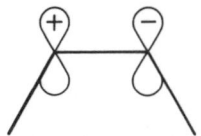

whereas, after the conrotatory rotation, the σ* orbital appears to be

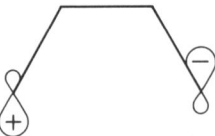

Notice that these orbitals have the same nodal patterns (phase relations) as the two occupied molecular orbitals of the product 1,3-butadiene:

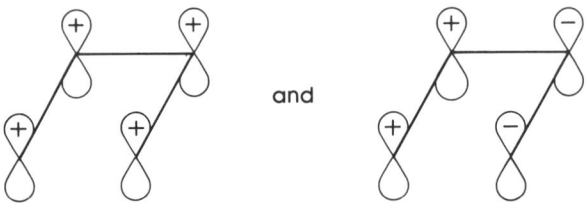

Thus, the same excitations that lead to rupture of the reactant bond also form the product bond. We shall make use of the idea of charge flow from the highest occupied molecular orbital (HOMO) to the lowest unoccupied molecular orbital (LUMO) again. In still other examples, it will be seen that orbitals other then the HOMO and LUMO can play important roles. What is important is that low-energy single excitations are present that permit new bonds to form as old bonds break—this is the essence of the concerted reactions we are studying.

4.7. Electrocyclic Reactions by Occupied-Orbital Following

Another feature of chemical reactions can be seen by again considering the opening of a four-membered ring but with a heteroatom present. The (hypothetical) opening of

to give

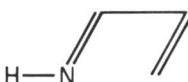

has the same orbital-symmetry constraints as the reaction examined in previous sections. However, symmetry is missing, so the idea of rigorous orbital *symmetry* cannot be used. Furthermore, orbital symmetry cannot be the cause of the barrier to disrotatory opening of cyclobutene—the true origin of symmetry barriers is the noncorrelation of *nodal patterns* among the reactant and product orbitals (Pearson, 1976). Such nodal-pattern noncorrelation is examined in this section.

By applying a disrotatory motion to the two *occupied* orbitals of

one obtains

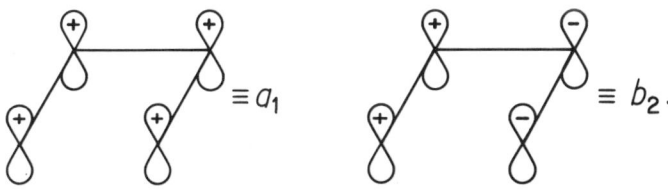

[in which DIS(σ) denotes the effect of the disrotatory opening on the σ orbital]. In the product

the occupied π orbitals look like

Notice that although DIS(σ) has the same nodal pattern (phase relationship) between the atomic orbitals as occurs in orbital a_1, the DIS(π_{23}) orbital does not match in its phase relationship with b_2. Thus, such ring opening is forbidden. It is important to keep in mind that we are examining the effect of the DIS opening on the *occupied* reactant orbitals to see whether the *occupied* product orbitals result—exactly what is done when constructing an orbital-correlation diagram. That is, orbital crossings that take electrons from occupied reactant orbitals into unoccupied product orbitals may arise. When this occurs, the reaction is forbidden by symmetry considerations.

On the other hand, a conrotatory (CON) motion applied to the occupied active orbitals of cyclobutene gives

$$\text{CON}(\sigma) \equiv$$

and CON(π_{23}) is identical to DIS(π_{23}). The phases of CON(π_{23}) and CON(σ) agree with those of the product a_1 and b_2 orbitals, respectively, so for this motion the *occupied* orbitals correlate. For disrotatory motion they do not.

This kind of orbital-following works equally well on the hypothetical reaction involving aza-substituted cyclobutene. The phase relationships of the *occupied* reactant orbitals are followed as the molecular deformation of interest takes place. Only the *occupied* molecular orbitals need to be considered because of the bond-symmetry rule discussed in section 4.4. By ignoring the unoccupied molecular orbitals, work is reduced, but the opportunity is lost to guess how large a symmetry barrier is expected, because to know how steeply uphill the configuration-correlation diagram should be drawn, the relative energies of the excited orbitals must also be known.

We now examine another electrocyclic reaction

and ask whether CON or DIS motion is allowed. Disrotatory motion is considered first. To simplify the diagrams in this and several of the following examples, the symbols + and − are used to indicate whether positive π-orbital components project up or down from the plane of the page. Thus, for disrotatory motion,

$$\text{DIS}(\sigma) \quad = \quad \equiv$$

$$\text{DIS}(\pi_1) \quad =$$

and

$$\text{DIS}(\pi_2) \quad =$$

The occupied π orbitals of the product are described by

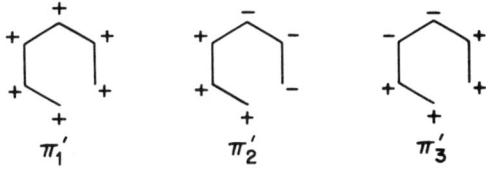

The orbitals can be correlated according to their phase relationships as

DIS(π_1) → π_1'

DIS(σ) → π_3'

DIS(π_2) → π_2',

so disrotatory ring opening is allowed. In contrast, for conrotatory motion,

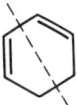

which could correlate with π_2' but then DIS(π_2) would have no product orbital with which to connect. Hence, conrotatory motion is forbidden. Note that the motion (CON or DIS) that is allowed varies, depending upon the length of the conjugated π network connecting the two termini (Woodward and Hoffman, 1970).

These electrocyclic reactions can also be treated by the HOMO-LUMO mixing concept. For example, in the disrotatory opening of

the reaction coordinate preserves the σ_v phase running through the molecule. As a result, the HOMO-LUMO single excitations ψ_k must have the same symmetry under σ_v as ψ_0 for $\langle \psi_0 | \partial h_e / \partial Q | \psi_k \rangle$ to be nonvanishing. This means that the HOMO and LUMO themselves must have the same symmetry under σ_v. The relevant HOMO's are

whereas the LUMO's are

Clearly σ and π^* have the same symmetry as π and σ^*, and the HOMO-LUMO excitations give rise to patterns that allow the new bonds to form. Thus, this reaction is also allowed according to the LUMO-HOMO mixing criterion.

4.8. Cycloaddition Reactions by Orbital Following

The cycloaddition reaction

$$2\ H_2C = CH_2 \rightarrow \square$$

considered in section 4.4 is now reexamined by using the orbital-following procedure. The fragmentation F of the two σ bonds in

$$\square$$

gives

$$F\sigma_1 = F\ \{\text{diagram}\} = \{\text{diagram}\}\ \text{or}\ \{\text{diagram}\}$$

$$F\sigma_2 = F\ \{\text{diagram}\} = \{\text{diagram}\}\ \text{or}\ \{\text{diagram}\}.$$

Here, for example, $F\sigma_1$ is used to represent the result of fragmentation on the σ_1 orbital.

$F\sigma_1$ has the same phase properties as the bonding π orbital of one ethylene fragment. However, $F\sigma_2$ looks like an antibonding π^* orbital of the other ethylene fragment, so this fragmentation is forbidden, in agreement with the results of section 4.4.

EXAMPLES FOR GROUND-STATE THERMAL PROCESSES   59

Notice that the symmetry-combined σ orbitals, σ_1 and σ_2, were used because orbitals were needed that, on bond rupture, would have amplitude on *both* atoms of the resultant ethylene molecule. If we had considered

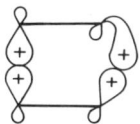

we would achieve no *phase* information, because two or more pieces of a wavefunction or orbital are needed to make a statement about phase.

The above cycloaddition reaction is labeled $[2_s + 2_s]$. The 2's denote the conjugation lengths of both fragments; the subscripts s show that fragments react in a suprafacial manner. The term *suprafacial* refers to an attack on the same face of the π system; *antarafacial* (subscript a) means that the newly formed or broken bonds occur on opposite faces of the π system.

We now consider the $[2_s + 2_a]$ cycloaddition of two ethylenes. To indicate the fact that the one ethylene is bonded in a suprafacial manner and the other in an antarafacial manner, cyclobutane is drawn as follows:

in which the + and − signs again indicate the directions of the orbital lobes participating in the bonding. The fragmentation results in the following orbital mappings:

The fragmented orbitals with the lowest-energy nodal pattern that results have been chosen to correlate. For example,

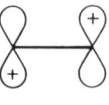

is not drawn in the first case. These fragmented orbitals have the proper phase relationships to correlate with the bonding π orbitals of both ethylenes; hence, the $[2_s + 2_a]$ reaction is allowed.

The retrograde Diels-Alder reaction

$$\text{(cyclohexene peroxide)} \rightarrow \text{(butadiene)} + O_2$$

is a more difficult example. The fragmentation process gives rise to the following orbital mappings:

[orbital diagrams]

These three fragmented orbitals have the same phase patterns as in the three occupied active orbitals of

$$\text{(butadiene)} + O_2$$

hence, the retrograde Diels-Alder reaction is allowed via a $[4_s + 2_s]$ mechanism. Notice that the lengths of the conjugated π systems in the two ene

EXAMPLES FOR GROUND-STATE THERMAL PROCESSES

systems determine whether the cycloaddition reaction is allowed via the supra-supra mechanism (Woodward and Hoffman, 1970).

The above reaction can also be examined by the HOMO-LUMO method. The relevant HOMO's of the σ and π moieties are

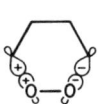 

and the LUMO's are

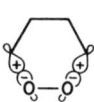 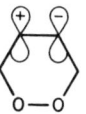

As the fragmentation of the two carbon-oxygen bonds begins, the σ_{CO} and π^* can be combined to form a new π bond in the diene

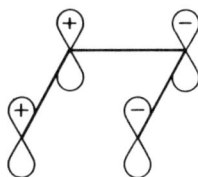

and σ^*_{CO} and π can combine to give the π bond on O_2

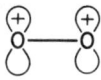

Thus, the $[4_s + 2_s]$ reaction is allowed.

The fragment HOMO's and LUMO's could also be chosen to refer to the O_2 and

product molecules. In this case, the O_2 HOMO

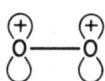

has good overlap with the diene LUMO

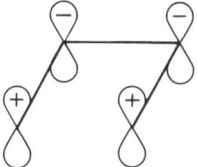

and the O₂ LUMO

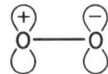

overlaps favorably with the diene HOMO

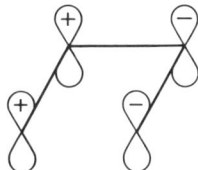

again leading to the prediction that the Diels-Alder reaction is symmetry-allowed.

In section 4.4 the four-center concerted-addition reaction typified by the dimerization of ethylene to give cyclobutane was shown to be thermally forbidden. However, in some circumstances, products can still be formed. For example,

can dimerize to give 1,5-cyclooctadiene

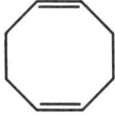

What makes this reaction occur so easily (i.e., with low activation energy), when the analogous ethylene dimerization has a very large activation barrier, is the availability of a *two-step* reaction path. By first combining two reactant molecules to give the radical intermediate

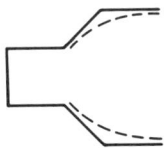

and, in a subsequent step, coupling the two allyl radicals, one obtains

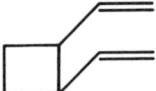

which can then rearrange via an *allowed* Cope rearrangement (Pearson, 1976) to give the final product 1,5-cyclooctadiene. The stability of the well-known allyl radical intermediates make this reaction path thermally favored. This path does not contribute to the ethylene dimerization because the 1,4-biradical

is very unstable. In contrast, perfluoroethylene F_2CCF_2 can thermally dimerize via the radical pathway because of the strong radical-stabilizing influence of the electron-withdrawing fluorine atoms. These examples illustrate the point made in the introduction to Part 2. The symmetry analysis can be applied to any *single* reaction step, but if that step is predicted to be forbidden, one must keep in mind that other pathways might be available and even favorable.

4.9. Sigmatropic Migrations via HOMO-LUMO Overlaps

The reaction

utilizes the [1,5] suprafacial shift of a hydrogen atom. If the product were

if would be a [1,5] antarafacial migration. This class of reaction can be examined by considering HOMO-LUMO charge-flow interactions (i.e., a singly excited configuration interaction caused by $\langle \psi_0 | \partial h_e / \partial Q | \psi_k \rangle$). The HOMO's of the two relevant fragments are

$$\pi = \qquad \qquad \sigma =$$

and the LUMO's are

$\pi^* =$  $\sigma^* =$

As the hydrogen atom is moved from the right carbon terminus to the left terminus, both the σ and π^* orbitals and the π and σ^* orbitals develop substantial σ- and π-bonding overlap. If σ^* were drawn

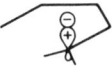

the singly excited configuration $\sigma^2\pi\sigma^*$ would still give rise to bonding interactions, because this configuration would simply enter into the perturbation description of ψ_k (or $\delta\rho_{k0}$) with an opposite sign (relative to the sign it has with σ^* drawn as originally drawn.) Another way to say this is that

are equivalent descriptions of the same orbital. An *overall* sign change never has any influence on the physical content of an orbital (or a wavefunction)—it is only the internal *relative* phases that are important.

Based on this analysis, the [1,5] suprafacial shift is allowed according to the criterion of HOMO-LUMO overlap criterion. The antarafacial shift is forbidden because the orbital of the hydrogen atom attaches to the bottom of the p_π orbital of the left carbon terminus. As a result, the σ and π^* orbitals and the π and σ^* orbitals no longer have *totally* favorable overlap for the product molecules (they have some favorable and some unfavorable overlap):

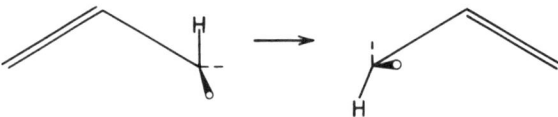

In contrast, the [1,3] antarafacial migration

is allowed because the relevant HOMO's

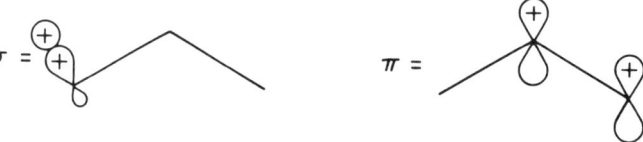

and LUMO's

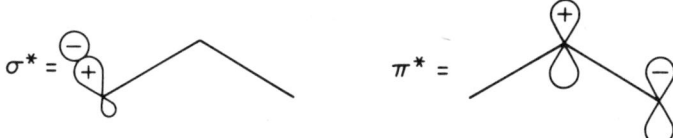

develop good overlap ($\sigma \leftrightarrow \pi^*$ and $\pi \leftrightarrow \sigma^*$) as the hydrogen moves to the bottom of the left terminus. The [1,3] suprafacial shift is forbidden because the feature that determines which reaction (antara or supra) is allowed is the phase pattern of the π orbital of the reactant (Woodward and Hoffmann, 1970).

Fukui (1971) has extended the idea of good fragment HOMO-LUMO overlap to predict *where* in a molecule the reaction is most likely to occur. For example, in the reaction

$$Cl-CH_3 + Cl^- \rightarrow Cl^- + H_3C-Cl$$

the HOMO of the attacking Cl^- is

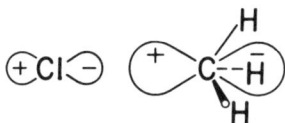

and the LUMO of $ClCH_3$ is

The other HOMO-LUMO pair is very high in energy because it involves the excited state of Cl^-; hence, it contributes little to the charge flow. Attack of the Cl^- should occur either on the back end of the $ClCH_3$, where the LUMO has large amplitude, to give $Cl^- + H_3CCl$, or on the Cl end of $ClCH_3$ to give $ClCl + {}^-CH_3$. The former reaction is thermodynamically favored because of the differences in bond strengths ($C^-Cl > Cl^-Cl$) and electron affinity ($Cl \gg CH_3$). The fundamental point in using the Fukui method is to identify the HOMO and LUMO pairs of the fragments and then to let them interact in a way (i.e., along some molecular distortion) that allows them to overlap *maximally*.

4.10. A Topology-Based Method

In this section, a final technique for predicting whether a reaction is allowed is described. This method, as developed by Zimmerman (1966) and by Dewar (1966), concentrates on the energetics of a *proposed* transition state. In this approach, the energy is governed by the aromatic or antiaromatic nature of the transition state. By concentrating on aromaticity, this tool is limited to reactions that pass through a *cyclic* transition state—a class of reactions called *pericyclic* (Woodward and Hoffman, 1970). The primary advantage of this technique is that it requires knowledge of only the *topology* of the transition state and of the number of active electrons—it is not necessary to examine individual molecular orbitals, HOMO's, or LUMO's.

To implement this method, one begins by assigning phases to the atomic orbitals involved in the cyclic transition state in order to give positive overlaps—as far as possible—as one walks along the bonds being broken and the bonds being formed throughout the transition state. It may not be possible to have positive overlap throughout, for one or more *interorbital* sign inversions may be forced by the nature of the *atomic* orbitals being used. If an odd number of such sign inversions occurs, the transition state is said to be *Möbius*; no inversions or an even number of inversions give rise to a *Hückel* transition state. A Hückel transition state is said to be stable if it contains $4n + 2$ electrons and unstable if it contains $4n$; Möbius transition states are stable if they contain $4n$ electrons.

Let us use this method to determine whether the reaction Mg $(3s^2)$ + $H_2 \rightarrow MgH_2$ $(^1A_1)$ is allowed. For C_{2v} symmetry, $3s_{Mg}$ is a_1 and $H_2 \sigma_g$ is a_1, but the MgH σ bonds are a_1 and b_2; thus, the bond-symmetry rule indicates that the reaction is forbidden. The HOMO-LUMO overlap criterion yields the same result, namely,

$$H_2 \sigma_u = \text{LUMO} \quad \text{and} \quad 3s_{Mg} = \text{HOMO}.$$

In the Dewar-Zimmerman topology-based method the atomic orbitals are assigned phases as follows to permit no sign changes:

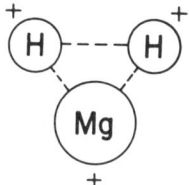

This is a Hückel system containing $4n = 4$ active electrons; hence, its transition state is unstable and the reaction is forbidden. In contrast,

$$\text{Ni } (4s^2 3d^{10}) + H_2 \rightarrow NiH_2$$

is allowed *if* Ni uses its *d* orbitals, since the system is now Möbius,

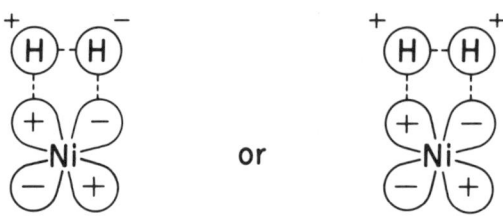

and contains $4n = 4$ *active* electrons (only the two electrons from H_2 and the two electrons from the one active d orbital count.)

The suprafacial addition of H_2 to H_2CCH_2 is a 4-electron Hückel system.

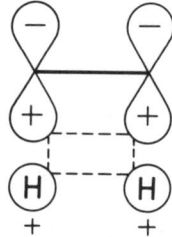

It is forbidden, just as would be found via a LUMO-HOMO or full-symmetry-correlation treatment. The antarafacial addition reaction is allowed, since it has a four-electron Möbius transition state. The antarafacial attack diagram can be constructed by walking along the new-bond–old-bond cycle as follows (starting with the top hydrogen and moving to the neighboring carbon)

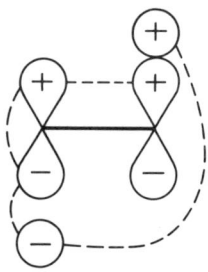

which shows one phase change, or alternatively, by walking from the top hydrogen to the bottom hydrogen

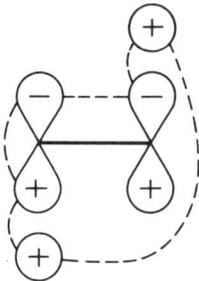

where again one *inter*orbital phase change results. Hence, this reaction is Möbius and contains four electrons. Note that this method applies only to those systems having (proposed) transition states that are cyclic. Other methods that we have talked about (in particular, full orbital diagrams, configuration-correlation diagrams, and HOMO-LUMO interactions) can be used to solve any problem as long as some symmetry exists or as long as one can identify the proper HOMO and LUMO of the fragment.

It would be good practice for the reader to go back through sections 4.4–4.9 and apply all of the Dewar-Zimmerman, HOMO-LUMO, symmetry-correlation, and orbital-following techniques. This would help one to see relationships among the methods. In particular, it is important to observe that most of the techniques that we have covered are closely interrelated. Each attempts to determine whether the *occupied* orbitals of the reactant (where electronic *configuration* information appears) evolve smoothly into *occupied* orbitals of the products. In the methods of the orbital-correlation diagram and the bond-symmetry rule, this is achieved by matching symmetries of occupied orbitals. In the orbital-following technique, one concentrates on phase relations of these orbitals. The HOMO-LUMO method monitors the evolution of the occupied orbitals by looking for low-energy virtual orbitals that can be mixed with the occupied orbitals of the reactant to generate product-occupied orbitals. Although the Dewar-Zimmerman method emphasizes the aromatic or nonaromatic nature of the *cyclic* transition state, even these properties are directly related to the occupied-orbital nodal patterns.

Problems

1. Consider the reaction

The products could be those shown, or the cyclohexane could have the opposite stereochemistry, that is,

Which of these reactions is thermally allowed? Use HOMO-LUMO, the bond-symmetry rule, and Zimmerman-Dewar methods.

2. Which of the ketones shown below should be thermally unstable with respect to CO loss to give the diene and cyclotriene, respectively?

3. Use orbital-correlation diagrams, configuration-correlation diagrams (using the two mirror planes of symmetry), and orbital-following to predict whether the reaction given below is thermally allowed.

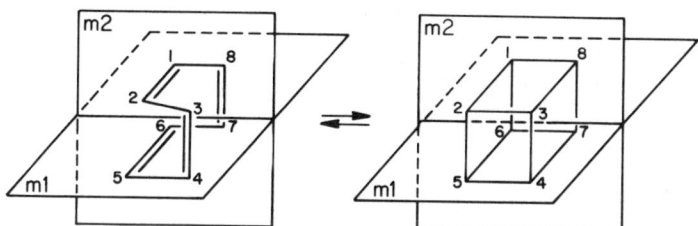

4. An all-cis polyene

$$\underset{D}{\overset{F}{\diagdown}}C=C(-C=C)_n-\underset{R_1}{\overset{H}{C}}\cdots R_2$$

is known to undergo a thermal sigmatropic hydrogen shift to give a 50:50 mixture of two compounds that have identical chemical formulas but differ in their geometrical and absolute configurations. What are the two products if n is even? What are they if n is odd? You may assume that steric factors are unimportant in formulating your answer. Use the Zimmerman-Dewar method to answer this problem.

5. Even though the [1,3] suprafacial hydrogen shift

is forbidden, the corresponding methyl-group migration is allowed. Why? What products (including absolute and geometrical isomer considerations) are expected in the [1,3] suprafacial migration of the R_5, R_6, R_7 substituted methyl group in

You may neglect steric factors.

6. Ammonia NH_3 is known to have C_{3v} symmetry. The bonding in this molecule can be explained by using the $2s$ and $2p$ orbitals of nitrogen and the $1s$ orbital of hydrogen.

   a. Use symmetry projectors to combine the $1s_H$ orbitals to form symmetry-adapted orbitals. What are the symmetries of the resulting orbitals?

   b. Repeat step (a) for the $2s_N$ and three $2p_N$ orbitals.

   c. Show in a qualitative orbital-energy-level diagram how the N and H atomic orbitals combine to produce bonding, nonbonding, and antibonding molecular orbitals of NH_3. Label the orbitals. Assume that the $1s_H$ orbitals are lower in energy than the $2p_N$ orbital but higher than the $2s_N$ orbital.

   For use in the following questions, *seven* molecular orbitals are labeled in order of increasing energy $\phi_1 \ldots \phi_7$.

   d. Consider the singlet excited state that can be generated by exciting an electron from ϕ_3 into the lowest available molecular orbital. Is this excited state stable, first-order Jahn-Teller unstable, or second-order Jahn-Teller unstable? If it is Jahn-Teller unstable, what symmetry of vibration would be expected to distort the molecule (NH_3 has $2A_1$ and $2E$ vibrations)?

   e. Repeat step (d) for an excitation out of ϕ_3 into the highest-available-valence molecular orbital.

Part 3
Theory and Applications Pertaining to Photochemical Processes

Chapters 1–4 treated thermal reactions. In Chapters 5 and 6 the theoretical concepts needed to understand photochemical reactions are explained. Chapter 5 begins by reviewing the qualitative experimental features that characterize most photochemical processes. The purpose of this review is to focus attention on those aspects of particular excited-state potential energy surfaces that play crucial roles in rate-determining processes because the slow rate-determining steps determine the outcome of photochemical reactions. An important principle is the Kasha rule (Kasha, 1950) that radiationless transitions among excited singlet states *usually* occur very quickly compared either to fluorescence or to a chemical reaction. Thus, it is only necessary to consider the lowest-excited-singlet-state potential energy surface, since it is on this surface that the system may eventually undergo chemical reaction. Chapter 5 also shows that the strategy for using symmetry concepts to probe photochemical problems is very much the same as for thermal reactions. It is still necessary to seek symmetry-imposed activation barriers in the appropriate (excited) potential energy surface. In addition, avoided (or real) crossings of the excited- and ground-state surfaces must be sought, since the system yields the commonly observed ground-state products through these funnels.

Chapter 6 presents a quantitative analysis of the rates of internal conversion and intersystem crossing. This treatment will provide understanding of how such rates depend upon the shapes and energy spacings of the potential energy surfaces between which the radiationless process takes place, though the treatment does not provide the mechanics to compute the rates from first principles. This analysis will show that radiationless rates will be rapid when the surfaces approach one another closely with similar slopes and when there are high frequency vibrational modes available for digesting the excess electronic energy. In the case of intersystem crossing, considerations of orbital angular momentum favor certain transitions [e.g., $^1(\pi\pi^*) \rightarrow\, ^3(n\pi^*)$] over others [$^1(n\pi^*) \rightarrow\, ^3(n\pi^*)$]. Knowing how these radiationless rates depend upon physical characteristics of the ground and excited states of the molecule enables one to design molecules with particular photochemical behavior and to interpret experimental data from many photoreactions.

In Chapter 7 the principles introduced in the Chapters 5 and 6 will be applied in a qualitative manner to several photoreactions.

Readers not entirely familiar with the Franck-Condon principle and other concepts relating to photon-absorption processes should read Appendix B before beginning Chapters 5-7.

Chapter 5

Introductory Remarks about Photochemical Reactions

5.1. Nature of Low-Energy Excited States

Electronic excitation arising from the absorption of one or more photons usually gives rise to an excited state having the same *spin symmetry* as the absorbing (ground) state. Transitions that do not conserve spin are more likely when the absorbing molecule contains one or more heavy atoms that allow the various spin states to be mixed via spin-orbit coupling (Turro, 1978). Excited states of different spin symmetry can also be populated via collisional energy transfer from an excited sensitizer molecule or atom (e.g., benzophenone or mercury).

The vast majority of stable compounds that can easily be used as reaction-starting materials have closed-shell singlet ground states (S_0). Therefore, for most of the material in this chapter we assume that the ground state is a singlet. However, most of the conclusions are valid also for molecules having other types of ground states.

One-electron excitations arising from one-photon transitions can give rise to either singlet ($S_1, S_2, \ldots$) or triplet ($T_1, T_2, \ldots$) excited states. By convention these states are ordered on an electronic energy basis when assigning the labels S_1, T_1, and so on. Usually the triplet state arising from a given orbital transition (e.g., $n\pi^*$ or $\pi\pi^*$) lies below its corresponding singlet state. The most common explanation for this arrangement is that the triplet state contains two electrons of the same m_s value and this lowers the energy by exchange interaction relative to the singlet—that is, the two electrons of the triplet state are kept apart by the Pauli exclusion principle. However, this explanation is not complete. For excited states in which a bonding-to-antibonding orbital transition occurs, the charge distribution character of the singlet and triplet states can differ significantly, and this difference in electron distribution must also be considered.

Singlet $(\pi\pi^*)^1$ excited states can be represented approximately by Slater determinant wavefunctions of the form

$$\frac{1}{\sqrt{2}} [|\ldots \pi\alpha\pi^*\beta| - |\ldots \pi\beta\pi^*\alpha|] \qquad (5.1)$$

(Cook, 1978; Pilar, 1968; Turro, 1978). Using the following (approximate) representations of the π and π^* orbitals in terms of the individual atomic orbitals (P_A and P_B) and the amplitudes (x and y) on each atom,

$$\pi \cong xP_A + yP_B \qquad (5.2)$$

$$\pi^* \cong yP_A - xP_B, \qquad (5.3)$$

to express the *active electron parts* of the Slater determinants in terms of their atomic components, we find

$$(\pi\pi^*)^1 \cong \{2xy[P_AP_A - P_BP_B] + (y^2 - x^2)[P_BP_A - P_AP_B]\} \frac{(\alpha\beta - \beta\alpha)}{\sqrt{2}}. \qquad (5.4)$$

The same analysis of the triplet state determinant gives

$$(\pi\pi^*)^3 = |\ldots \pi\alpha\pi^*\alpha| = [(x^2 + y^2)(P_BP_A - P_AP_B)] \frac{\alpha\alpha}{\sqrt{2}}. \qquad (5.5)$$

In writing equations 5.4 and 5.5, the last two columns of appropriate Slater determinants are simply expanded; the other columns contain the *passive* orbitals, which need not be explicitly addressed here. For π bonds that are not extremely polar $x \cong y$, in which case the $(\pi\pi^*)^1$ state is dominated by the *ionic* terms $(P_AP_A - P_BP_B)$, whereas the $(\pi\pi^*)^3$ state contains only radical or *covalent* terms $(P_BP_A - P_AP_B)$. This strong difference in the two charge densities plays an important role in making the $\pi\pi^*$ triplet state lower in energy than the singlet (which has both electrons in the same region of space).

The description of the relative energetics of the singlet and triplet states just given has been simplified by assuming that the π and π^* orbitals can be expressed in terms of the same atomic orbitals. More sophisticated *ab initio* calculations on singlet $\pi\pi^*$ states indicate that it is more correct to think of the π and π^* orbitals as having different radial extent or size (McMurchie and Davidson, 1977). However, detailed investigations of this problem indicate that the formulation in terms of the ionic and covalent charge-density picture is qualitatively correct. It is important to keep in mind that, as a result of the large difference in electron distribution, these excited states may have physical and chemical properties that are very different from those of the ground state and that depend upon spin multiplicity (Pearson, 1976; Turro, 1978).

5.2. Energy Redistribution in the Singlet Manifold

We now consider what happens to the energy that has been stored in the electronic framework of the system after an excited state is formed. In an isolated-molecule gas-phase system (Rice, 1971) the total energy must be conserved within the degrees of freedom of the electronic-and-nuclear motion (vibration, rotation, translation) of the molecule. In condensed media, it is possible eventually (on time scales appropriate to vibrational relaxation, $\sim 10^{-12}$–10^{-10} sec) for energy to be dissipated to the surrounding medium.

Two experimental observations indicate the processes that occur with the highest rate (those fast enough to lead to observed phenomena). First, it is observed that fluorescence ($S_n \rightarrow S_0 + h\nu$) almost never occurs from higher singlet states (S_n, $n > 1$) (however, see Beer and Longuet-Higgins, 1955). Even when S_2, S_3, ... are excited from S_0, the fluorescence usually comes from S_1, a fact known as Kasha's rule (1950). This rule states that when S_2, S_3 ... are excited, some of this *electronic* energy must be digested (on a time scale that is fast with respect to fluorescence, namely, $< 10^{-6}$–10^{-10} sec) by the degrees of freedom associated with nuclear motion, since in this isolated molecule the *total* energy must remain constant. This process of digesting electronic energy ($S_{n+1} \rightarrow S_n$) is called *internal conversion*; its physical origin will be explored in more detail in the following chapter.

A second important observation is that the products of most photochemical reactions of molecules that contain more than a few (~ 5) atoms need *not* be electronically excited. Thermal and photochemically induced chemiluminescent reactions do occur, but they are exceptions. Most photochemical reactions yield directly (without much fluorescence) a high fraction of the products in their *ground* (S_0) state. This experimental fact indicates that photoreactions do not occur entirely on an excited-state potential energy surface (most likely, S_1) followed by radiative decay (fluorescence) of the products; rather, the excited-state reactants often end up on the ground-state potential energy surface of the product molecule(s). How then can an excited-state reaction ever be symmetry-forbidden if the ground-state energy of the products lies below the energy of the photochemically prepared state? To answer this question it is necessary to understand how internal conversion occurs. A more detailed analysis of internal conversion will show that, if the S_0 and S_1 potential energy surfaces vary with some reaction coordinate (geometrical distortion) as in Figure 5-1, then the initially excited S_1 molecules will have a good chance to react and to yield ground-state (S_0) products by hopping from the S_1 surface to S_0 near the *close approach* of S_0 and S_1. In Chapter 6 it will be shown that the rate of hopping depends on the number of internal degrees of freedom that the molecule can use to digest the electronic energy. Near the

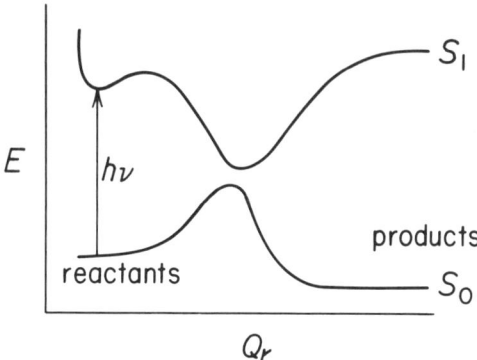

Figure 5-1
A typical arrangement of S_0 and S_1 surfaces that results in an allowed excited-state reaction.

avoided crossing, there is little electronic energy to be digested, so internal conversion is facilitated at these geometries. For small molecules with low densities of internal states (vibration, rotation, translation) the rate of hopping is slower and the system is more likely to remain on S_1. Remember that, in gas-phase radiationless transitions, total energy must be conserved. Therefore, the system gains internal energy even though the diagram in Figure 5-1 shows the system hopping from S_1 to S_0, thereby losing electronic energy. This internal energy can be vibrational motion along Q_r, or it can be vibration or rotation along degrees of freedom orthogonal to Q_r (not shown in the figure). The larger the number of vibrational modes, the less likely it is for the energy to remain for a substantial length of time in the Q_r degrees of freedom. In condensed media, the excess vibrational-rotational energy can also be dissipated to surroundings in 10^{-10}–10^{-12} sec.

On the other hand, if the S_0 and S_1 surfaces have the properties shown in Figure 5-2, the initially prepared S_1 molecules will have to overcome an *additional* activation barrier for reaction to occur. Such S_1 surfaces would then lead to photochemically forbidden reactions: the excited molecule will be trapped on the reactant side of the barrier. If the photon energy placed the system high enough on the S_1 surface to overcome the barrier, the S_1 molecule may undergo internal conversion to S_0 near an avoided crossing or actual intersection of S_0 and S_1. As will be seen later, the excited S_1 molecules can most efficiently hop (undergo internal conversion) to S_0 at such near crossings of S_0 and S_1 (Michl, 1972, 1974, 1975). After reaching the S_0 surface, the system can either yield products or the reactants can be restored. The quantum yields for these two processes will depend upon both the precise shape of the S_1 and S_0 surfaces and the photon energy. The nature of the distortion coordinates that lead to such *funnels* determines the reaction product. After a *likely* reaction

INTRODUCTORY REMARKS ABOUT PHOTOCHEMICAL REACTIONS

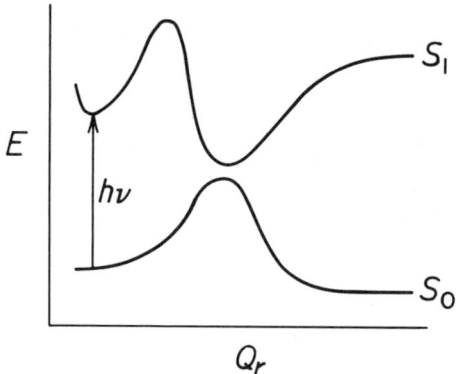

Figure 5-2
A typical arrangement of S_0 and S_1 surfaces that results in a forbidden excited-state reaction.

coordinate is obtained, it is necessary to seek symmetry-imposed reaction barriers that arise on the S_1 surface between the reactant geometry (as determined by the Franck-Condon principle; see Herzberg, 1966, and Appendix B) and the funnel geometry through which the reaction proceeds. As in Chapter 1-3, this again means finding where the potential surface (S_1) passes through extrema (barriers and minima) as a result of avoided crossings.

Appendix B outlines the theory of an electronic transition in which a ground-state (S_0) molecule absorbs a photon and moves to an excited surface S_n $(n \geq 1)$. Both the common Franck-Condon approach and a partly classical picture are presented; these tools allow one to guess the geometry at which a molecule will enter an excited-state (S_n) potential energy surface. These entry geometries must be known before walking along the S_1 surface as outlined above can be undertaken.

5.3. Processes Involving Triplet States

Thus far, our attention has been restricted to processes that take place on the singlet-state manifold. If spin-orbit effects were entirely negligible, the triplet states would not become populated after the primary photochemical event, and these spin-forbidden states could be ignored. However, triplet excited states are important because even a small amount of spin-orbit coupling can cause intersystem crossings $(S_n \rightarrow T_m)$, especially when the S_n and T_m surfaces approach or intersect. Moreover, the excited triplet states can also be directly populated in the primary event either by using electron impact on the neutral S_0 or by forming the doublet anion of S_0 (S_0^-) and subsequently detaching *an* electron from one of the orbitals occupied in S_0 (not from the LUMO of S_0 occupied by the extra electron of the anion).

When a triplet state becomes populated, it can undergo several processes. Higher triplet states (T_n, $n > 1$) usually decay in a radiationless manner to T_1 on a time scale that is very fast relative either to backward intersystem crossing —$T_1 \to S_1$ or S_0—or to radiative decay (phosphorescence)—$T_n \to S_m + h\nu$— both of which are slow spin-forbidden events. After undergoing internal conversion ($T_n \to T_1$), the T_1-state molecules usually live for 10^{-3}–10 sec. During this time, which is much longer than the radiative lifetime of the S_1 state (10^{-10}–10^{-6} sec), the triplet molecules can undergo chemical reaction or decay radiatively or by intersystem crossing to S_0. Since T_1 usually lies below S_1, it can no longer return to the excited singlet manifold. We will make this point clearer in the next chapter. The longer lifetime of the T_1 state plays an important role in its chemical reactivity—namely, time is available for reagents to arrange themselves in a geometrically favorable manner for initiation of reaction. The lifetime gives rise to one of the distinguishing features of triplet molecules—namely, the tendency to undergo two-step *radical reactions* rather than the *concerted* one-step reactions characteristic of singlets. For a reagent in the triplet state to react with a closed-shell (singlet) molecule and yield a closed-shell product—for example,

$$O_2(^3\Sigma) + H_2CCH_2 \to H_2\overset{\overset{\displaystyle O-O}{\displaystyle ||}}{C-CH_2})$$

requires intersystem crossing. Such reactions are usually described by two steps. First, a triplet reagent bonds to the singlet molecule to form an intermediate triplet radical, for example,

$$H_2\overset{\overset{\displaystyle O-O\cdot}{\displaystyle |}}{C-\overset{\cdot}{C}H_2}$$

which slowly undergoes intersystem crossing to become a singlet radical that finally allows the two radical fragments to close and form a bond.

When an excited molecule in a singlet or triplet state reaches the lowest T_1 state, the factors that determine whether it will react to form products and the products that will be formed are the same as those used in analyzing the fate of the S_1-state molecules. The two preceding figures can be employed to explain the allowed and forbidden triplet reactions to yield singlet products by simply replacing S_1 by T_1 and making two modifications. First, the possibility for the T_1 and S_0 surfaces to intersect in a $(3N-7)$-dimensional space (for nonlinear molecules) must be considered; S_1 and S_0 can intersect in, at most, a $(3N-8)$-dimensional space. Second, when examining the rate of hopping from T_1 to S_0, spin-orbit coupling effects must be included. The rates of $S_1 \to S_0$ internal conversion and $T_1 \rightarrow S_0$ intersystem crossing are much different because the physical factors that govern these rates are different.

To predict whether a triplet-state reaction is allowed, one proceeds in the following way. First, the Franck-Condon principle is used to select those regions of nuclear configuration space that, upon the primary $(S_0 + h\nu \rightarrow S_n)$ event, are likely to be populated. The internal conversion $(S_n \rightarrow S_1)$ then occurs rapidly to give S_1. Then, likely molecular deformation coordinates are sought, along which S_1 and T_1 approach closely or intersect and along which no high symmetry-imposed barriers on the S_1 surface occur. The intersystem crossing $(S_1 \rightarrow T_1)$ funnels will determine the molecular geometries at which T_1 is formed. Finally, starting at this point on the potential energy hypersurface at which T_1 is formed, additional deformations are sought, such that T_1 and S_0 intersect or approach closely and along which no high barriers exist on the T_1 surface. These $T_1 \rightarrow S_0$ funnels can yield either ground-state reactants or ground-state products, depending on the detailed nature of the $T_1 \rightarrow S_0$ funnel or intersection.

Note that the approach in studying photochemical processes is similar to that for thermal reactions. Minima in S_1 and T_1 are sought because these minima often indicate where the excited surfaces (S_1 and T_1) come closest to S_0. Although it is indeed possible for S_1 and T_1 to *intersect* S_0, emphasis should not be placed on these crossing geometries; they contribute a subspace of smaller dimension than the dimension of S_0, T_1, or S_1. The regions of space in which S_1 or T_1 comes close (within striking distance for the non-Born-Oppenheimer terms in the total Hamiltonian) are of higher dimension and thus more important. In addition to finding minima in S_1 or T_1, the points at which S_1 and T_1 can cross or come close must be determined, and possible barriers in S_1 and T_1 must be sought.

Before proceeding to examples of photochemical reactions, it is necessary to explore further the physical mechanisms by which radiationless processes such as internal conversion and intersystem crossing occur. This knowledge is important because it will permit us both to predict when these events will be likely (occur at competitive rates) and to understand how isotopic substitution, heavy atoms, and vibrational state densities can be used to alter the rates at which they occur. This is the subject of Chapter 6.

Chapter 6

Internal Conversion and Intersystem Crossing

6.1. The States Between Which Transitions Occur

To understand the mechanisms by which a molecule can undergo a *radiationless transition* (Yardley, 1980; Lin, 1980) from one potential energy surface to another, the Schrödinger equation for combined electronic and nuclear motion given in Chapter 1 is needed. The electronic wavefunctions $\{\phi_a(\mathbf{r}|\mathbf{R})\}$ corresponding to the two interacting states between which transitions occur obey the equations

$$h_e \phi_{S_0} = E_{S_0}(\mathbf{R}) \phi_{S_0} \tag{6.1}$$

and

$$h_e \phi_x = E_x(\mathbf{R}) \phi_x \qquad (x = S_1 \text{ or } T_1). \tag{6.2}$$

Within the Born-Oppenheimer approximation, the internal (vibrational-rotational) wavefunctions belonging to the S_0 and excited potential surfaces obey the equations

$$(D_R^2 + E_{S_0})\chi_v^0 = \epsilon_v^0 \chi_v^0 \tag{6.3}$$

and

$$(D_R^2 + E_x)\chi_{v'}^x = \epsilon_{v'}^x \chi_{v'}^x, \tag{6.4}$$

in which D_R^2 is the kinetic energy operator for all of the nuclear vibration and rotation. The energies ϵ_v^0 and $\epsilon_{v'}^x$ are the *total* Born-Oppenheimer energies of $\phi_{S_0}\chi_v^0$ and $\phi_x\chi_{v'}^x$, respectively. ϵ_v^0 can be decomposed into the electronic energy at the minimum of the S_0 surface plus the χ_v^0 internal energy (e_v^0)

$$\epsilon_v^0 = E_{S_0}(\text{min}) + e_v^0. \tag{6.5}$$

An analogous expression can be written for ϵ_v^x

$$\epsilon_v^x = E_x(\min) + e_v^x \tag{6.6}$$

where $E_x(\min)$ is the electronic energy at the minimum of the excited-state surface. $E_x(\min) - E_{S_0}(\min)$ gives the adiabatic *electronic* energy difference for the $S_0 \rightarrow X$ excitation; e_v^0 and e_v^x are simply the vibration/rotation energies (labeled by the quantum number v) on the S_0 and x surfaces, respectively.

In the approximation that the internal vibrations and rotations may be uncoupled, the functions χ_v^0, and $\chi_{v'}^x$ consist of products of appropriate rotational functions and of $3N - 6$ vibrational wavefunctions—one for each of the normal or local vibrational coordinates (Yardley, 1980) *including* the reaction coordinate Q_r. As pointed out in Chapter 1, motion along coordinates orthogonal to Q_r can often be thought of as involving approximately harmonic vibration. However, the components of χ_v^0 and $\chi_{v'}^x$ that describe motion along Q_r cannot be approximated by harmonic motion except near local minima. In regions of Q_r space in which S_0 has negative curvature, the Q_r component of χ_v^0 looks like a continuum wavefunction rather than a bound vibrational wavefunction.

We now consider the transitions used when a molecule hops from S_1 or T_1 to S_0. The S_1 or T_1 state has been populated by the mechanism $S_0 + h\nu \rightarrow S_n \rightarrow (S_1, T_1)$. In the Born-Oppenheimer approximation, the wavefunction of this excited state is given by

$$\psi_x = \phi_x(\mathbf{r}|\mathbf{R})\chi_{v'}^x(\mathbf{R}). \tag{6.7}$$

The vibrational energy level $\epsilon_{v'}^x$ may be high or quite low (e.g., in condensed-phase situations). Although $\epsilon_{v'}^x$ also contains rotational and, perhaps, relative translational energy, we will, for brevity, speak of this energy as being vibrational. If the density of states (states per cm^{-1} of energy) in the S_0 manifold is high at this energy level ($\epsilon_{v'}^x$), it is likely that there is a state of the S_0 manifold

$$\psi_0 = \phi_{S_0}(\mathbf{r}|\mathbf{R})\chi_v^0(\mathbf{R}) \tag{6.8}$$

that is *nearly degenerate* with ψ_x. These two zeroth-order states will be coupled by the terms in the true Hamiltonian that give rise to non-Born-Oppenheimer corrections. This coupling will be strong if the off-diagonal matrix elements $\langle \psi_x | H | \psi_0 \rangle$ are nonnegligible when compared to the energy difference $\epsilon_{v'}^x - \epsilon_v^0$ (Yardley, 1980). Therefore, in this situation the non-Born-Oppenheimer coupling is said to give rise to *transitions* between ψ_x and ψ_0, and these transitions are the hopping that has been discussed. If the excited state is a triplet, the non-Born-Oppenheimer terms alone would not couple ψ_x and ψ_0; H must contain spin-orbit-type interactions if $\langle \psi_x | H | \psi_0 \rangle$ is to be nonvanishing.

6.2. Rates of Transitions

To evaluate *rates* of such transitions, the conventional Fermi "golden rule" can be used (Yardley, 1980; Lin, 1980); this rule states that transitions starting in $\phi_x \chi^x_{v'}$ and going to $\phi_{S_0} \chi^0_v$ and caused by the non-Born-Oppenheimer parts of $H - h_e$ occur at a rate given in $\sec^{-1}$ by

$$W = \frac{2\pi}{\hbar} \sum_v |\langle \phi_{S_0} \chi^0_v | H - h_e | \phi_x \chi^x_{v'} \rangle|^2 \delta(\epsilon^0_v - \epsilon^x_{v'}). \tag{6.9}$$

The δ function guarantees that the states $\phi_{S_0} \chi^0_v$ contributing to the total radiationless transition rate have the same Born-Oppenheimer energy as $\phi_x \chi^x_{v'}$. When many vibrational or rotational modes are present, there may be many χ^0_v functions, each having the same energy ϵ^0_v. The number of such states is referred to as the density of states ρ at this total energy (ϵ^0_v)

$$\rho(\epsilon^x_{v'}) = \sum_v \delta(\epsilon^0_v - \epsilon^x_{v'}). \tag{6.10}$$

If there is reason to believe that all of the states $\{\phi_{S_0} \chi^0_v\}$ in this degenerate manifold couple to the same extent with the initial state $\phi_x \chi^x_{v'}$, then the sum over v in the above expression for W can be replaced by the appropriate state density

$$W = \frac{2\pi}{\hbar} |\langle \phi_{S_0} \chi^0_v | H - h_e | \phi_x \chi^x_{v'} \rangle|^2 \rho(\epsilon^0_v) \tag{6.11}$$

in which $\chi^x_{v'}$ is any one of the degenerate states. Modern research on the behavior of electronically excited molecules indicates that, even for systems with high state densities, often only a small fraction of the modes play an active role in the radiationless transition. As a result, it may not be wise to use equation 6.11 when trying to understand radiationless transition rates; it is probably more appropriate to think in terms of equation 6.9.

Internal Conversion Rates

We now consider how equation 6.9 depends on the electronic energy difference and the vibrational wavefunctions χ^0_v and $\chi^x_{v'}$ for the internal conversion case in which ϕ_x is a singlet state. The terms in $H - h_e$ that couple the initial state $\phi_{S_1} \chi^x_{v'}$ to the final state $\phi_{S_0} \chi^0_v$ are the non-Born-Oppenheimer terms, and the off-diagonal coupling matrix element described in Chapter 1 is

$$\langle \phi_{S_0} \chi^0_v | H - h_e | \phi_{S_1} \chi^x_{v'} \rangle = \langle \phi_{S_0} \chi^0_v | \chi^x_{v'} D^2_R \phi_{S_1} + 2 D_R \phi_{S_1} \cdot D_R \chi^x_{v'} \rangle. \tag{6.12}$$

As discussed in Chapter 1, the second term in this expression is usually larger than the first (except when the second term vanishes, owing to symmetry); hence, the analysis will proceed using only this term. The analysis of the first term can be performed in analogous fashion (Berry, 1966); to do so would not shed further light on the physical origins of internal conversion.

An expression for the above coupling matrix element that is more physically useful can be obtained by applying the D_R operator to the Born-Oppenheimer Schrödinger equation, which ϕ_{S_1} obeys:

$$D_R(h_e \phi_{S_1} - E_{S_1} \phi_{S_1}) = 0. \tag{6.13}$$

Multiplying on the left by ϕ_{S_0} and integrating over the *electronic* coordinates only yields

$$\langle \phi_{S_0} | D_R h_e | \phi_{S_1} \rangle + \langle \phi_{S_0} | h_e D_R \phi_{S_1} \rangle - D_R E_{S_1} \langle \phi_{S_0} | \phi_{S_1} \rangle$$

$$- E_{S_1} \langle \phi_{S_0} | D_R \phi_{S_1} \rangle = 0. \tag{6.14}$$

Equation 6.14 can be solved for $\langle \phi_{S_1} | D_R \phi_{S_1} \rangle$, which can then be used to reexpress the coupling matrix element and, therefore, to write the transition rate as

$$W = \frac{2\pi}{\hbar} \sum_v \delta(\epsilon_v^0 - \epsilon_{v'}^0)$$

$$|\langle \chi_v^0 | [E_{S_1} - E_{S_0}]^{-1} \langle \phi_{S_0} | D_R h_e | \phi_{S_1} \rangle \cdot 2 D_R \chi_{v'}^x \rangle|^2 \tag{6.15}$$

This analysis shows that W is likely to be large if regions of nuclear configuration space (**R**) exist for which the electronic energy gap is small ($E_{S_1}(\mathbf{R}) \cong E_{S_0}(\mathbf{R})$) and the product $\chi_v^0(\mathbf{R}) D_R \chi_{v'}^x(\mathbf{R})$ is nonvanishing. Therefore, molecular deformations that bring the two singlet-state potential surfaces close to one another should be sought. If χ_v^0 and $D_R \chi_{v'}^x$ have appreciable *overlap* in the region in which $E_{S_0} \cong E_{S_1}$, then internal conversion is likely. However, the *electronic* force matrix element $\langle \phi_{S_0} | D_R h_e | \phi_{S_1} \rangle$ must also be substantial; this integral will be large if (in the orbital-following sense introduced in Chapter 4) the distortion tends to evolve the orbital structure of ϕ_{S_1} into that of ϕ_{S_0}. Where symmetry is present, the direct product of the S_0 and S_1 symmetries must match that of $D_R h_e$. For example, in H_2CO, the $n\pi^*$ state is achieved by excitation from an occupied $b_2(n)$ orbital to the vacant $b_1(\pi^*)$ orbital. The kind of motion that is symmetry-consistent with $b_1 \times b_2$ is a_2. H_2CO does not have any vibration with a_2 symmetry. As a result, the $\langle \phi_{S_0} | D_R h_e | \phi_{S_1} \rangle$ matrix element should be small for $n\pi^*$ states in C_{2v} symmetry. In contrast, $\pi\pi^*$ S_1 states have $b_1 \times b_1 = a_1$ symmetry and, hence, D_R can be a_1 because S_0 is also a_1. H_2CO has three a_1 vibrations (symmetric CH stretch, CO stretch, and HCH bend) of which the CO stretch would be ex-

INTERNAL CONVERSION AND INTERSYSTEM CROSSING 87

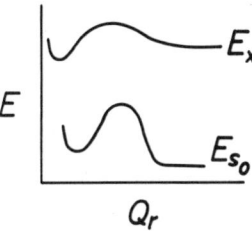

Figure 6-1
Energy surfaces that do not approach closely.

pected to have the largest effect on mapping the π^* orbital into the π orbital because the nature of these two orbitals is affected by the distance from C to O.

Let us review the procedure developed so far. If a coordinate exists along which E_{S_0} approaches E_{S_1}, the hopping rate for internal conversion is increased. Near this avoided crossing or near approach of the S_0 and S_1 surfaces, one considers whether symmetry or physical force are likely to make $\langle \phi_{S_0} | D_R h_e | \phi_{S_1} \rangle$ significant for deformations either along Q_r ($D_R = D_{Q_r}$) or along some direction perpendicular to Q_r. Directions in which both this electronic force matrix element is large and the vibrational product $\chi_\nu^0 D_R \chi_{\nu'}^x$ is substantial will play important roles as modes that *digest* the excess electronic energy $E_{S_0} - E_{S_1}$. That is, the direction along which E_{S_0} and E_{S_1} come close is important because this motion brings the molecule to the funnel geometry. Once the molecule is near the funnel, it can use *other* degrees of freedom (orthogonal to Q_r) to digest the excess electronic energy.

Energy-Digesting Modes

Two extreme cases of how the E_{S_0} and E_x surfaces might appear may be distinguished. The first case pertains to situations in which E_{S_0} and E_x do not approach one another closely—in other words, within an energy gap that is approximately equal to a non-Born-Oppenheimer matrix element. The shapes of two such surfaces are shown in Figure 6-1. Efficient internal conversion may still be possible by transfer of electronic energy to internal vibrational energy. To analyze the rates of such processes in this case, note first that the energy-denominator factor in equation 6.15—$[E_{S_0}(Q) - E_{S_1}(Q)]^{-1}$—is small and never undergoes rapid growth near some critical geometry as it would, for example, in the near-crossing situations (a second special case that will be treated shortly). Therefore, $E_{S_0} - E_{S_1}$ is approximated as a part that depends on the reaction coordinate Q_r, plus a part that describes the (approximately harmonic) motion perpendicular to Q_r:

$$E_{S_0} - E_{S_1} \cong E_{S_0}(Q_r) - E_{S_1}(Q_r) + E_{S_0}(\mathbf{R}) - E_{S_1}(\mathbf{R}). \tag{6.16}$$

Furthermore, the shapes of $E_{S_0}(\mathbf{R})$ and $E_{S_1}(\mathbf{R})$ are assumed to be sufficiently similar that $E_{S_0}(\mathbf{R}) - E_S(\mathbf{R})$ can be neglected relative to the presumed large value of $E_{S_0}(Q_r) - E_{S_1}(Q_r)$ splitting.

To continue the analysis, two additional assumptions are made: (1) Only one mode, whose spatial coordinate is R_a, plays an active role as an energy acceptor (Yardley, 1980). (2) In the initial S_1 state, the active mode is in its $v' = 0$ level. Assumption 2 is by no means fully justified or even necessary. In condensed media situations, it might be more justified, since vibrational energy would probably have been dissipated to the surroundings prior to the internal conversion process. This assumption is made only so the resulting integral containing $\chi_v^0(R_a)$ and $D_{R_a}\chi_{v'}^x(R_a)$ can be physically interpreted more easily. We will then argue that essentially the same physical picture would be obtained, after more tedious algebraic manipulation, if the more general ($v' \neq 0$) case were analyzed (see Yardley, 1980). Under the above outlined limitations, the rate expression (equation 6.15) becomes

$$W = \frac{2\pi}{\hbar} \sum_v \delta(\epsilon_v^0 - \epsilon_{v'}^x) |\langle \chi_v^0(Q_r)[E_{S_0}(Q_r) - E_{S_1}(Q_r)]^{-1} \chi_{v'}^x(Q_r)\rangle$$

$$\left(\prod_{b \neq a} \langle \chi_{v_b}^0 | \chi_{v_b'}^x \rangle\right) \langle \phi_{S_0} | D_{R_a} h_e | \phi_{S_1} \rangle_{eq} 2 \langle \chi_{v_a}^0 | D_{R_a} | \chi_{v_a'}^x \rangle|^2. \quad (6.17)$$

Here the product Π_b extends over all modes *other than* Q_r and the active R_a and gives rise to simple Franck-Condon overlap factors for the passive modes. If these modes are fully passive, then the shapes of the S_0 and S_1 surfaces along these directions should be identical, in which case the $\langle \chi_{v_b}^0 | \chi_{v_b'}^x \rangle$ overlap factors would reduce to products of simple δ-functions $\Pi_b \, \delta_{v_b,v_b'}$. In writing the above expression for W, we assumed that the electronic force matrix element is rather insensitive to R_a. Consequently, $\langle \phi_{S_0} | D_{R_a} h_e | \phi_{S_1} \rangle$ was evaluated at the equilibrium value of the S_1 state of R_a ($R_a = R_a^{eq}$), which was denoted by the subscript "eq" in this integral. The vibrational quantum numbers appearing in ϵ_v^0 and $\epsilon_{v'}^x$ are underlined because they contain the quantum numbers of *all* modes (for example, $\mathbf{v} = v_{Q_r}, v_a, v_b; b = 1 \ldots$).

With all of these observations, W reduces to

$$W = \frac{2\pi}{\hbar} \sum_{v_a v_r v_r'} \delta(e_{v_a}^0 + E_{S_0}(\min) + e_{v_r}^0 - e_{v_a'}^x - E_{S_1}(\min) - e_{v_r'}^x)$$

$$|\langle \chi_{v_r}^0(Q_r) | [E_{S_0}(Q_r) - E_{S_1}(Q_r)]^{-1} \chi_{v_r'}^x(Q_r)\rangle|^2$$

$$|\langle \phi_{S_0} | D_{R_a} h_e | \phi_{S_1} \rangle_{eq}|^2 4 |\langle \chi_{v_a}^0 | D_{R_a} | \chi_{v_a'}^x \rangle|^2 \quad (6.18)$$

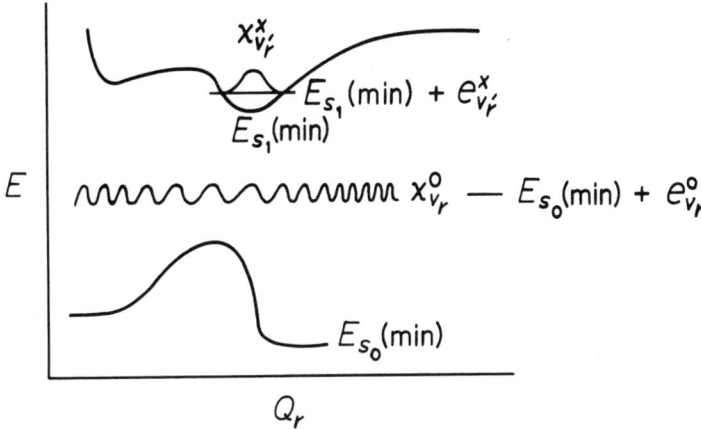

Figure 6-2
Energies relevant to internal conversion.

in which $e_{v_r}^0$ and $e_{v_r'}^x$ are the *internal* energies in the Q_r mode in the two states. The energy-conservation δ-function requires that the excess energy

$$E_{S_0}(\min) - E_{S_1}(\min) + e_{v_r}^0 - e_{v_r'}^x \equiv \Delta E_r,$$

which is equal to an adiabatic electronic energy difference plus the amount of energy *along* Q_r that has to be "digested," is balanced by the change in vibrational energy of the energy-accepting mode. This modified energy gap is shown in Figure 6-2.

If $\chi_{v_a}^0$ and $\chi_{v_a'}^x$ are approximated by simple harmonic oscillator functions having identical frequencies ω but equilibrium bond lengths that differ by ΔR_a, the evaluation of the $\langle \chi_{v_a}^0 | D_{R_a} | \chi_{v_a'}^x \rangle$ integral is straightforward. For the case of $v_a' = 0$ (Yardley, 1980)

$$\langle \chi_{v_a}^0 | D_{R_a} | \chi_0^x \rangle = \sqrt{\frac{\omega}{\hbar}} \left[\sqrt{\frac{v_a}{2}} (-\sqrt{X})^{v_a-1}((v_a-1)!)^{-1/2} - \sqrt{\frac{v_a+1}{2}} (-\sqrt{X})^{v_a+1}((v_a+1)!)^{-1/2} \right] \exp(-X/2) \quad (6.19)$$

is obtained, in which $X \equiv \omega/(2\hbar)\Delta R_a^2$. Substitution of this result into equation 6.18 yields the final expression for the rate of internal conversion:

90 CHAPTER 6

$$W = \frac{4\pi\omega}{\hbar^2} \sum_{v_r v_r'} |\langle \phi_{S_0} | D_{R_a} h_e | \phi_{S_1} \rangle_{eq}|^2$$

$$|\langle \chi_{v_r}^0(Q_r)[E_{S_0}(Q_r) - E_{S_1}(Q_r)]^{-1} \chi_{v_r'}^x(Q_r) \rangle|^2$$

$$e^{-X} \frac{\Delta E_r^2}{2\pi\hbar\omega(\Delta E_r - \hbar\omega)}^{1/2}$$

$$\exp\left[-\frac{\Delta E_r - \hbar\omega}{\hbar\omega} \left(\ln\left(\frac{\Delta E_r - \hbar\omega}{\hbar\omega X} \right) - 1 \right) \right] \tag{6.20}$$

in which ΔE_r is the *excess energy* or gap defined earlier.

Notice that ΔE_r depends upon $e_{v_r}^0 - e_{v_r'}^x$, the energy change along the mode Q_r. Since for this special case $E_{S_0}(Q_r) - E_{S_1}(Q_r)$ does not become small for any value of Q_r, the element $\langle \chi_{v_r}^0 | (E_{S_0} - E_{S_1})^{-1} | \chi_{v_r'}^x \rangle$ is likely to be small unless $\chi_{v_r}^0$ and $\chi_{v_r'}^x$ have very similar shapes. This will *not* be the case if $\Delta E_r \cong 0$, since then $\chi_{v_r}^0$ would correspond to a wave packet having high kinetic energy and a short de Broglie wavelength (as in Figure 6-2). This function could have little overlap with any low-energy $\chi_{v_r'}^x$. Hence, the dominant contribution is for $v_r' = 0$ and small v_r, and therefore it is reasonable to set $v_r' = v_r = 0$ in equation 6.20.

Notice that equation 6.20 leads to the conclusion that *high frequency vibrational modes* should be most effective in digesting the excess energy. For such modes, $(\Delta E_r - \hbar\omega)/\hbar\omega$ is as small as possible. The *exponential dependence* of W on the energy gap ΔE_r is thought to give rise to the observations leading to the Kasha rule. For most molecules the $S_0 - S_1$ spacing (near the S_0 equilibrium geometry populated in the Franck-Condon absorption process) is larger than the $S_1 - S_2$, $S_2 - S_3$, ... splittings. Hence, internal conversion from S_1 to S_0 is slower than between higher states, since ΔE_r is larger for the $S_1 \rightarrow S_0$ transition.

Equation 6.20 also shows that if experimentally one desired to modify the rate of internal conversion by isotopic substitution, the high frequency variations should be modified. For example, substitution of deuterium for hydrogen should produce substantial changes in the rate of internal conversion.

In the other extreme case (see Figure 6-3) in which $E_{S_0}(Q) - E_{S_1}(Q)$ becomes small along Q_r (of the order of magnitude of the non-Born-Oppenheimer matrix elements), we assume for motion along directions perpendicular to Q_r that $E_{S_0} - E_{S_1}$ can be written as two components—one consisting of motion along Q_r and a second comprised of harmonic segments (having the same geometries and frequencies). These harmonic potentials are assumed to be identical on S_0 and S_1 and, hence, cancel yielding

$$E_{S_0} - E_{S_1} \cong E_{S_0}(Q_r) - E_{S_1}(Q_r), \tag{6.21}$$

INTERNAL CONVERSION AND INTERSYSTEM CROSSING

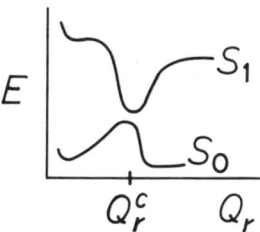

Figure 6-3
Energy surfaces that do approach closely.

In this case, Q_r itself is the energy-digesting mode.

The treatment in this section of the special case in which E_{S_0} and E_{S_1} do not approach closely can also be made for this case in Figure 6-3 to the point at which

$$W = \frac{2\pi}{\hbar} \sum_{v_r} \delta(E_{S_0}(\min) + e_{v_r}^0 - E_{S_1}(\min) - e_{v_r'}^x)$$

$$4|\langle \chi_{v_r}^0|(E_{S_0}(Q_r) - E_{S_1}(Q_r))^{-1} \langle \phi_{S_0}|D_{Q_r}h_e|\phi_{S_1}\rangle D_R|\chi_{v_r'}^x\rangle|^2 \quad (6.22)$$

Because $(E_{S_0} - E_{S_1})^{-1}$ enhances contributions to the integral over Q_r near the point of closest-approach Q_r^c, the electronic force matrix element and the energy difference can be approximated by their values at Q_r^c to obtain

$$W = \frac{8\pi}{\hbar} \sum_{v_r} \delta(E_{S_0}(\min) + e_{v_r}^0 - E_{S_1}(\min) - e_{v_r'}^x)$$

$$|[E_{S_0}(Q_r^c) - E_{S_1}(Q_r^c)]^{-2}|\langle \phi_{S_0}|D_{Q_r}h_e|\phi_{S_1}\rangle|^2 \langle \chi_{v_r}^0|D_{Q_r}|\chi_{v_r'}^x\rangle|^2 \quad (6.23)$$

Unfortunately, an energy-gap law is not easily obtained for this case because the $\langle \chi_{v_r}^0|D_{Q_r}|\chi_{v_r'}^x\rangle$ integral does not include two bound harmonic oscillator functions; $\chi_{v_r}^0$ describes *free* (unbound) motion along the S_0 surface. However, transitions will be *favored* if the electronic energy gap $E_{S_0}(Q_r^c) - E_{S_1}(Q_r^c)$ is small and the electronic-force matrix element is large. Moreover, the *kinetic energy* of motion along Q_r in the product S_0 state is $E_{S_1}(\min) + e_{n_r'}^x - E_{S_0}(Q_r^c)$, at $Q = Q_r^c$.

The Landau-Zener Point of View

Before discussing the rates of intersystem crossing, it is useful to point out the connection between the treatment just given and the Landau-Zener method for looking at rates of surface hopping (Eyring, Walter, and Kimball, 1944), a method that pertains *only* to the second case treated above (the close approach of S_0 and S_1 depicted in Figure 6-3). In the Landau-Zener approach the near crossing of the S_0 and S_1 surfaces is parameterized by the slopes (f_0 and f_1) of the surfaces near their avoided crossing and the closest-approach energy $2\epsilon_{10} \equiv 2(E_{S_1}(Q_r^c) - E_{S_0}(Q_r^c))$. The probability of a surface hop (per vibration along Q_r, as dictated by χ_v^x) is then expressed as

$$P = 1 - \exp[-(4\pi^2 \epsilon_{10}^2)(hv|f_0 - f_1|)^{-1}] \qquad (6.24)$$

in which v is the velocity of the nuclei as they pass through the avoided-crossing region. The dependence of P on the vibrational level ($\chi_{v'}^x$) of the initial S_1 state comes from this velocity—if v_r' is large, the velocity is high. Equation 6.24 shows that three things—a close approach (small ϵ_{10}), fast-moving nuclei (high frequency vibration), and a small difference in slope (small change in force)—favor surface hopping. This influence of the change in slope (which is the change in the *forces* felt by the nuclear framework of the molecule along the Q_r direction) is obscured somewhat in the earlier expression for W (equation 6.23). This force effect is contained in the $\chi_{v_r}^0(Q_r) D_Q \chi_{v_r'}^x(Q_r)$ factor. If S_0 and S_1 have very different slopes near Q_r^c, the wavefunctions $\chi_{v_r}^0$ and $\chi_{v_r'}^x$ will have greatly different *local* de Broglie wavelengths in this region, and $\chi_{v_r}^0$ and $\chi_{v_r'}^x$ are not likely to have large local overlap. In contrast, similar slopes of S_0 and S_1 near Q_r^c will lead to large overlap of $\chi_{v_r}^0$ and $\chi_{v_r'}^x$ (i.e., similar shapes in $\chi_{v_r}^0$ and $\chi_{v_r'}^x$).

6.3. Intersystem Crossing Rates

In intersystem crossing rates the electronic wavefunctions ϕ_0 and ϕ_x are singlet and triplet, respectively. However, the spin-orbital operator

$$h_{SO} = \frac{e^2 \hbar}{2m^2 c^2} \left\{ \sum_{i,a} \frac{Z_a}{r_{ia}^3} (\mathbf{r}_{ia} \times \mathbf{P}_i) \cdot \mathbf{S}_i + \sum_{i \neq j} \left[\frac{(2\mathbf{P}_j - \mathbf{P}_i) \times \mathbf{r}_{ij}}{r_{ij}^3} \right] \cdot \mathbf{S}_i \right\} \qquad (6.25)$$

couples ϕ_0 and ϕ_x to give *perturbed* wavefunctions $\tilde{\phi}_0$ and $\tilde{\phi}_x$ that contain both singlet and triplet components. Intersystem crossing is viewed as occurring between these perturbed functions by a mechanism similar to that just discussed for internal conversion. The perturbed electronic wavefunctions are approximated by

$$\tilde{\phi}_{S_0} = \phi_{S_0} + \langle \phi_{S_0} | h_{SO} | \phi_{T_1} \rangle (E_{S_0} - E_{T_1})^{-1} \phi_{T_1} \qquad (6.26)$$

and

$$\tilde{\phi}_{T_1} = \tilde{\phi}_{T_1} + \langle \phi_{T_1}|h_{SO}|\phi_{S_0}\rangle (E_{T_1} - E_{S_0})^{-1}\phi_{S_0} \qquad (6.27)$$

(Eyring, Walter, and Kimball, 1944; Pilar, 1968). These functions can now be used in the Fermi golden-rule formula to evaluate W for intersystem crossing. In doing so, the electronic-force matrix element (or even the second non-Born-Oppenheimer factor $\langle \tilde{\phi}_{S_0}|D_Q^2 h_e|\tilde{\phi}_{T_1}\rangle$ that was not analyzed earlier) is modified because $\tilde{\phi}_{S_0}$ and $\tilde{\phi}_{T_1}$ are now of mixed spin character:

$$\langle \tilde{\phi}_{S_0}|D_Q h_e|\tilde{\phi}_{T_1}\rangle = \langle \phi_{S_0}|D_Q h_e|\phi_{T_1}\rangle$$

$$+\ \phi_{S_0}|D_Q h_e|\phi_{S_0}\ \phi_{T_1}|h_{SO}|\phi_{S_0}\ (E_{T_1} - E_{S_0})^{-1}$$

$$+\ \langle \phi_{S_0}|h_{SO}|\phi_{T_1}\rangle \langle \phi_{T_1}|D_Q h_e|\phi_{T_1}\rangle (E_{S_0} - E_{T_1})^{-1}$$

$$+\ \text{terms second order in } h_{SO}. \qquad (6.28)$$

The first term vanishes because of spin orthogonality since $D_Q h_e$ contains no spin-dependent terms. The other two terms contain electronic-force expectation values, $E_{S_0} - E_{T_1}$ energy denominators, and spin-orbit matrix elements. When squared and substituted into the expression for W (equation 6.23), these integrals give an expression for the rate of intersystem crossing. The treatment of *digesting modes* other than Q_r (if S_0 and T_1 remain far apart) and the treatment of the case of Q_r accepting the excess energy (when S_0 and T_1 cross or come very close) proceed in the same way as that for intersystem crossing; therefore, we need not repeat the analysis of the dependence of this radiationless rate on the energy gap, accepting-mode frequencies, and so forth.

The primary difference between the expressions for internal conversion (W_{IC}) and intersystem crossing (W_{ISC}) is contained in the spin-orbit integrals $\langle \phi_{S_0}|h_{SO}|\phi_{T_1}\rangle$ whose squares enter into W_{ISC}. These integrals require further discussion (Turro, 1978). The spin-orbit operator h_{SO}, consists of the dot product $(L_+ S_- + L_- S_+ + 2L_z S_z)$ of a *spatial electronic angular-momentum* operator and an *electric spin* operator. Clearly, it is the spin-operator components (S_+, S_-, but not S_z) that map the triplet spin function into the singlet spin function in the $\langle \phi_{S_0}|h_{SO}|\phi_{T_1}\rangle$ integral. When operating on ϕ_{T_1}, the corresponding spatial angular-momentum operator components can alter the angular characteristics of the spatial wavefunction in ϕ_{T_1}. More specifically, since the components L_x, L_y, L_z (or L_+, L_-, L_z) of the angular-momentum operator transform like rotation operators under point-group symmetry (see Appendix C or Cotton, 1963), the direct product of the spatial symmetries of ϕ_{S_0} and ϕ_{T_1} must match that of at least one of the components L_x, L_y or L_z for the spin-orbit matrix element to be nonvanishing. This is another important factor to keep in mind when deciding when intersystem crossing is likely to occur. For example, in H_2CO, the intersystem crossing transitions $(\pi\pi^*)^1 \to (n\pi^*)^3$ or $(n\pi^*)^3 \to n^2$ are spin-orbit *favored* because they utilize transitions between $\pi(b_1)$ and $n(b_2)$ orbitals or $\pi^*(b_1)$ and $n(b_2)$ orbitals whose direct product a_2 has the symmetry of a rotation about the symmetry axis of the molecule. Pictorially, this is represented by noting that a 90° *rotation* of the $n(b_2)$ orbital maps it into a π-like b_1 orbital. Likewise, the intersystem crossing rates $(\pi\pi^*)^1 \to (\pi\pi^*)^3$ and $(n\pi^*)^1 \to (n\pi^*)^3$ should be smaller because they are forbidden by first-order perturbation analysis —that is, the molecule has no rotation having $b_1 \times b_1 = a_1$ or $b_2 \times b_2 = a_1$ symmetry.

In the Landau-Zener method, which applies only to two surfaces (T_1 and S_0) that intersect or approach closely, the probability of intersystem crossing is given as before (equation 6.24) except that now the energy splitting $2\epsilon_{01}$ is *caused* by the spin-orbit coupling and is given by

$$\epsilon_{01} = \langle \phi_{S_0} | h_{SO} | \phi_{T_1} \rangle_{Q_f^c}. \tag{6.29}$$

Hence, the same conditions that favor internal conversion also favor intersystem crossing, except that the rate of intersystem crossing also includes the spin-orbit matrix element in a multiplicative manner. This element will be small unless heavy atoms are present and the two states can be connected by any of L_x, L_y, or L_z.

Chapter 7

Examples of Photochemical Reactions

In this chapter six photochemical reactions will be analyzed by the methods of Chapter 6.

7.1. Dimerization of Two Ethylenes

From the treatment of ground-state thermal reactions (Chapter 1–3) the four-center *concerted* dimerization of two S_0 ethylene molecules is forbidden by symmetry. Recall that in C_{2v} symmetry the orbital-correlation diagram for this reaction is that shown in Figure 7-1. If the ground state ($\pi^2\pi^2$, $\sigma^2\sigma^2$) and singly excited ($\pi^2\pi^*\pi$, $\sigma^2\sigma^*\sigma$) configurations of reactants and products and the configurations with which they correlate are included, the configuration-correlation diagram shown in Figure 7-2, in which the C_{2v} symmetries of these configurations are also indicated, is obtained. Not all of the configuration correlations are shown because it is important to examine first in more detail the type of photochemical event we wish to simulate. Following this examination, certain of these configurations can be eliminated.

Photochemical dimerization of ethylene might be viewed as a collision between $(\pi\pi^*)^1$ excited ethylene molecule and a ground-state π^2 ethylene under C_{2v} symmetry. Analysis of the reaction requires expressing this *localized* (non-symmetry-adapted) excited species in terms of symmetry orbitals. Using the *reverse* of the transformation from a localized orbital to a symmetry-adapted orbital, the proposed experimentally prepared $\pi_A^2(\pi\pi^*)_B^1$ state (in which A and B refer to the two isolated ethylene molecules) can be expressed as follows:

$$\pi_B = \pi_{a_1} + \pi_{b_2}, \qquad \pi_A = \pi_{a_1} - \pi_{b_2}$$

and

$$\pi_B^* = \pi_{b_1}^* + \pi_{a_2}^*, \qquad \pi_A^* = \pi_{b_1}^* - \pi_{a_2}^*$$

from which the localized (ethylene + ethylene*) configuration can be written

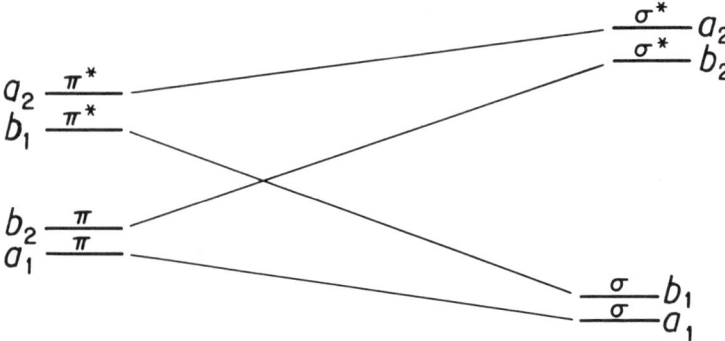

Figure 7-1
Orbital-correlation diagram for dimerization of ethylene.

$$\pi_A^2(\pi\pi^*)_B^1 = (\pi_{a_1} - \pi_{b_2})(\pi_{a_1} - \pi_{b_2})[(\pi_{a_1} + \pi_{b_2})(\pi_{b_1}^* + \pi_{a_2}^*)]^1$$

Keeping in mind that a singlet $(\pi\pi^*)^1$ state is represented by a combination of two Slater determinants, these orbital occupation patterns represent Slater determinant wavefunctions. This property of the wavefunctions, combined with the Pauli principle, allows one to eliminate the $\pi_{a_1}\pi_{a_1}\pi_{a_1}$ and $\pi_{b_2}^3$ pieces of this wavefunction; however, this local-orbital description still contains many symmetry pieces (e.g., $\pi_{a_1}^2\pi_{b_2}\pi_{b_1}^*$ has A_2 symmetry). The analysis of the various symmetry pieces of this localized function is analogous to decomposing the six ^2P states of Na* into four $^2\Pi$ and two $^2\Sigma$ states when Na* collides with H_2 in C_{2v} symmetry.

The goal of this analysis is to determine whether there exists at least one photochemically accessible path for the reaction. An answer could be obtained by first analyzing the symmetry elements that are in the $\pi_A^2(\pi\pi^*)_B^1$ state, then constructing *all* configuration-correlation diagrams consistent with *all* of these symmetries, and looking for symmetry-imposed barriers arising on *any* of these surfaces. This procedure would be exceedingly tedious but can be simplified by the following method.

The orbital configuration diagram in Figure 7-1 shows that an excitation from the $b_2\pi$ orbital to the $b_1\pi^*$ orbital would have the best chance of producing an *energetically downhill* reaction surface because the b_2 and b_1 orbitals undergo a crossing (which incidentally causes the thermal reaction to be forbidden). Certainly this $b_2 \rightarrow b_1$ excitation is a part of the local-orbital wavefunction described earlier (in that the $\pi_A^2(\pi\pi^*)_B^1$ function contains a $a_1^2b_2b_1$ component. This component is the one most likely to lead to stable products. Thus, a configuration-correlation diagram is constructed that contains only the ground-state (S_0) configurations (which have $^1A_1$ symmetry), the configurations to which they correlate, and the particular configurations having

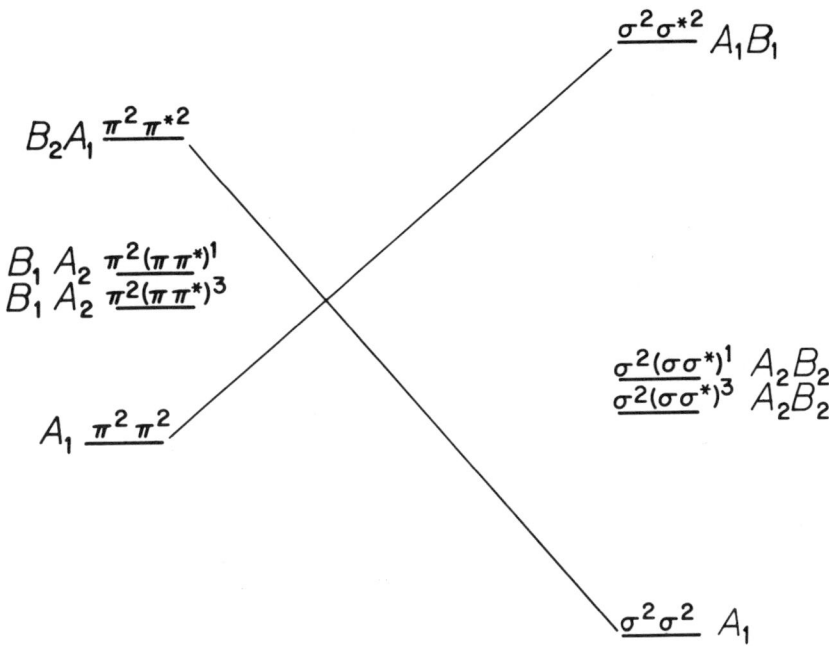

Figure 7-2
Configuration-correlation diagram for dimerization of ethylene.

$b_2 \times b_1 = A_2$ symmetry that arise from the $b_2 \to b_1$ excitation. Those configurations having B_1 and B_2 symmetry are eliminated from the configuration-correlation diagram in Figure 7-2. These eliminated configurations include $b_2 \to a_2$ and $a_1 \to b_1$ excitations that do not promote an electron from an orbital whose energy is increasing into an orbital whose energy is decreasing. The resulting simplified configuration-correlation diagram, which is *not* meant to be quantitatively accurate, is shown in Figure 7-3. If experimental data for the $\pi\pi^*$ and $\sigma\sigma^*$ excitation energies of ethylene and cyclobutane and for the thermochemical ΔE for this reaction were available, the configuration-correlation diagram could be made more quantitative.

Because of the (b_2, b_1) orbital crossing, the four configurations drawn in the configuration-correlation diagram would be degenerate at this crossing geometry at the level at which interelectron repulsion effects are ignored. However, the electron repulsions are not negligible and split the four configurations into four resulting *states*. At the crossing geometry, the triplet state is probably lowest in energy. The excited state, which is formed in the primary photochemical event, is—because of the forbidden nature of $S_0 + h\nu \to T_n$ events—likely to be a singlet state that involves a single orbital excitation because it was formed in a one-photon absorption. The $^1A_2\pi^2(\pi\pi^*)^1$ state fits

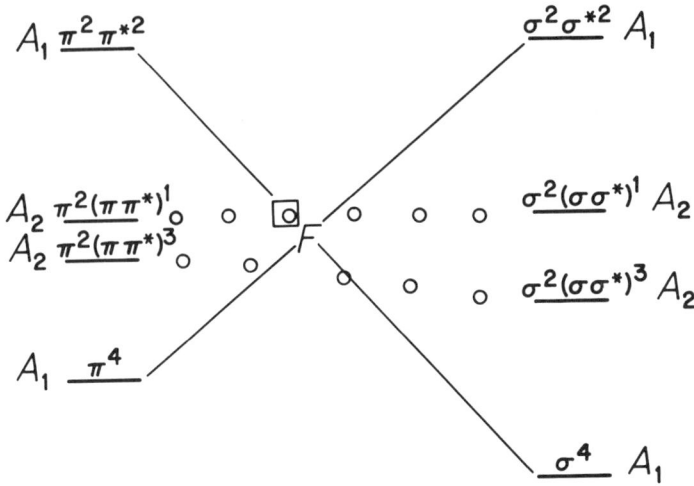

Figure 7-3
Simplified configuration-correlation diagram for dimerization of ethylene.

these requirements, though this state may only be the one most reactive component of the $\pi_A^2(\pi\pi^*)_B^1$ local-orbital excitation state.

After formation of the $^1A_2$ state, internal conversion can occur near the point enclosed in the box in Figure 7-3. As the two ethylenes collide, via an a_2 accepting-mode distortion, internal conversion to the *upper* $^1A_1$ state can be followed by "hopping" to the lower $^1A_1$ surface at the *funnel* (labeled F), thereby giving either reactant or product molecules in their $^1A_1\sigma^4$ *ground state*! The precise amount of ground-state reactant and $^1A_1$ and $^1A_2$ products depends upon the quantitative nature of the potential surfaces. Nowhere along the path from $^1A_2$ to $^1A_{1*}$ to $^1A_1$ is a reaction barrier encountered. The S_1 surface may be thermodynamically uphill if the $\sigma^2\sigma\sigma^{*1}$ state of cyclobutane lies above the $\pi^2\pi\pi^{*1}$ state of the ethylene dimer. In such a case, an appreciable quantum yield of product would not be expected until the *photon energy* used to populate the S_1 state is sufficient to exceed the energy gap between the $(\pi\pi^*)^1$ absorption threshold and the point at which the S_1 state intersects the upper $^1A_1$ curve. The crucial element in making this reaction photochemically allowed is the excitation of an electron from an orbital that moves uphill along Q_r to a downhill moving orbital. Notice that the purpose in constructing the configuration-correlation diagram was to see whether the system can efficiently (through internal conversion or funnels) get from the $^1A_2$ surface (at the Franck-Condon-populated *reactant* geometry) to the product ground-state surface without encountering any symmetry imposed barriers—*not* to follow the excited $^1A_2$ state all the way to products. Experimental evidence shows that internal conversion will be sufficiently fast that the product molecules need not be formed on the $^1A_2$ surface.

What would happen if the $^3A_2(\pi\pi^*)$ state were populated in the primary excitation event by triplet sensitization? If the configuration-correlation diagram in Figure 7-3 is quantitatively accurate, the T_1 surface moves downhill from reactants to products, and some phosphorescence from the $^3A_2(\sigma\sigma^*)^3$ state of cyclobutane is expected (assuming that the species does not first decompose to yield another product). In regions of Q_r space in which the $^3A_2$ surface intersects the *lower* $^1A_1$ surface (Figure 7-3), it is also possible that intersystem crossing could lead to the formation of ground-state *reactants*, because the lower $^1A_1$ surface is intersected on its reactant side. Recall that this intersystem crossing will be efficient only if one of the rotations R_x, R_y, and R_z has the same symmetry as the direct product $A_2 \times A_1 = a_2$. Because the C_{2v} point group does have a rotation with a_2 symmetry, intersystem crossing should be efficient. In summary, from the intersection of T_1 with S_0 on the left side, intersystem crossing should give ground-state reactant molecules. From the right-hand intersection of T_1 and S_0, products will form. Phosphorescence of products will also occur. The quantum yields for each of these three processes will depend upon the exact *values* of the radiative and intersystem crossing rate constants—such quantitative evaluations cannot be made from symmetry-based arguments.

7.2. Closure of 1,3-Butadiene to Cyclobutene

The orbital-correlation diagrams for the disrotatory (DIS) and conrotatory (CON) paths for closure of 1,3-butadiene are shown in Figure 7-4, in which e and o indicate even and odd symmetry under σ_v or C_2. In Chapter 4 the CON pathway for the thermal reactions was shown to be allowed and the DIS path to be forbidden. It will be seen in this section that the allowed pathway for the photochemical reaction differs. We begin by constructing a configuration-correlation diagram that includes only the *most important* configurations—namely, the ground state and the state most likely to lead to photoreaction. The fact that the energetically favorable $\pi \to \pi^*$ excitation has $o \to e$ symmetry for the DIS path is used next. For this reaction coordinate, the configuration-correlation diagram, which is not necessarily quantitatively accurate, is shown in Figure 7-5. If the relative-energy scale were correct in this figure—which depends on the strengths of the σ and π bonds and the strain energy of the cyclobutene—excitation of butadiene to the singlet $\pi^2 (\pi\pi^*)^1$ state would *not* yield cyclobutene at excitation energies near the $\pi\pi^{*1}$ threshold. Absorption of photons of higher energy might cause a reaction if the excess internal energy were maintained in the reaction coordinates. The reaction is *not* symmetry-forbidden, but the S_1 state of the butadiene must move uphill by a considerable

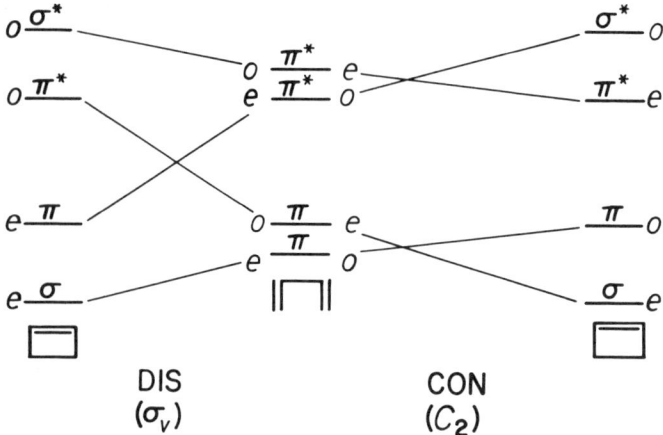

Figure 7-4
Orbital-correlation diagrams for CON and DIS closure of 1,3-butadiene.

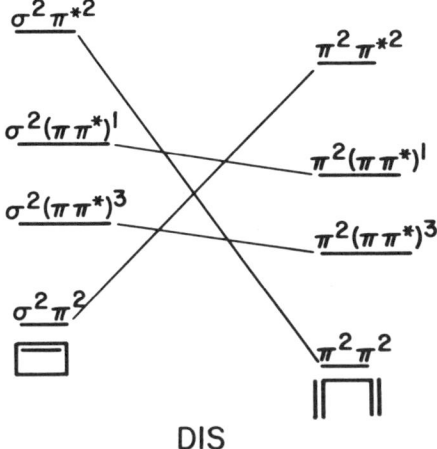

Figure 7-5
Configuration-correlation diagram for DIS closure.

amount to intersect with the upper totally symmetric surface. In contrast, the *reverse* S_1 photoreaction should occur readily because the $\pi^2(\pi\pi^*)^1$ state can cross to the *upper* totally symmetric ($\sigma^2\pi^{*2}$) surface and thereby permit funneling to the ground-state surface. The funneling can then produce either reactants or products in their ground state.

EXAMPLES OF PHOTOCHEMICAL REACTIONS   101

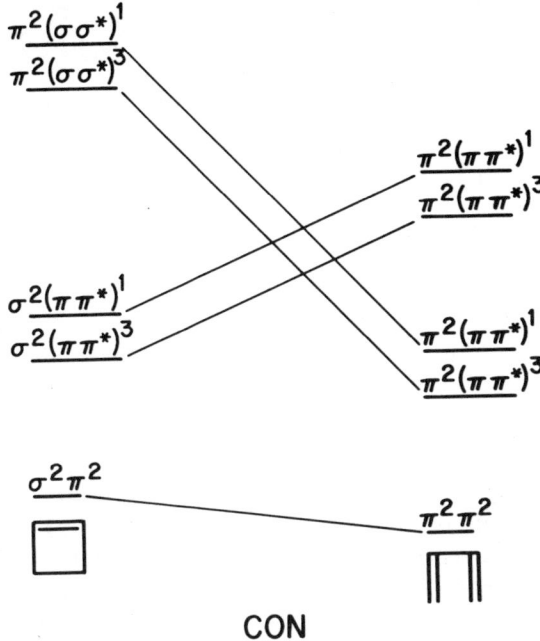

Figure 7-6
Configuration-correlation diagram for CON closure.

The CON reaction path, which is thermally allowed, is photochemically forbidden when the excitation includes promotion of an electron from the second orbital of the butadiene to the lowest π^* orbital. The relevant configuration-correlation diagram for the CON path is shown in Figure 7-6. It shows that in addition to the fact that reaction along the S_1 state of the butadiene to produce cyclobutene is endothermic—as was the case for the DIS process—an additional symmetry-imposed *barrier* to reaction on this S_1 surface is also present. Because no funnel route connecting S_1 to S_0 is present, it is unlikely that the system will return to S_0 at a geometry that characterizes cyclobutene; it is much more likely that the molecule will either fluoresce or return to S_0 via internal conversion near the reactant geometry, because it cannot move away from this geometry when it is on S_1. The thermal reaction is allowed in the CON case, as shown by the S_0 surface not moving uphill along Q_r. The same behavior of S_0 also makes the excited-state reaction to yield S_0 products forbidden. For the same reasons, excitation of the triplet $\pi^2(\pi\pi^*)^3$ butadiene should also fail to form cyclobutene.

The Dewar-Zimmerman rules can also be applied to certain photochemical reactions if the resonance stability rules are simply reversed (Pearson, 1976). That is, the DIS reaction has a $4n$-electron Hückel transition state

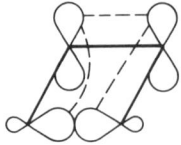

so this reaction is thermally forbidden and photochemically allowed. Analogously, the suprafacial [1,3] sigmatropic migration of a hydrogen atom is a 4-electron Hückel system

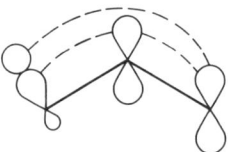

and would also occur photochemically. These simple rules *cannot* be universally valid because they do not contain references about the orbital into which the electron is excited. The range of applicability of the Dewar-Zimmerman prediction is limited to situations in which the occupied and virtual orbitals participating in the excitation are energy ordered, either as

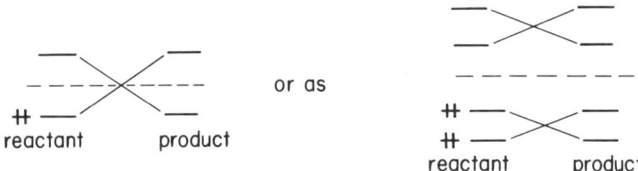

7.3. HOMO-LUMO (SOMO) Overlap for the Diels-Alder Reaction

In using the criterion of HOMO-LUMO overlap to study thermal reactions, the HOMO and LUMO orbitals for each of the two reactant species are examined. For a Diels-Alder reaction, the orbitals are the diene π_2 and π_3^* orbitals (Figure 7-7a) and the ene π and π^* orbitals (Figure 7-7b). The energy of the ordering of these orbitals and the ground-state reactant orbital occupations are depicted in Figure 7-7c. These diagrams show that $\pi_2 \rightarrow \pi^*$ and $\pi \rightarrow \pi_3$ excitations include favorable HOMO-LUMO interactions in that the low-energy excitations produce a charge flow that allows the *old bonds* to break as the *new bonds* form. Hence, the thermal Diels-Alder reaction is allowed.

To use the HOMO-LUMO overlap tool for the photochemical reaction, the orbitals are first occupied in a manner appropriate to the excited state. For example, to study the reaction of *excited* 1,3-butadiene with ethylene, the

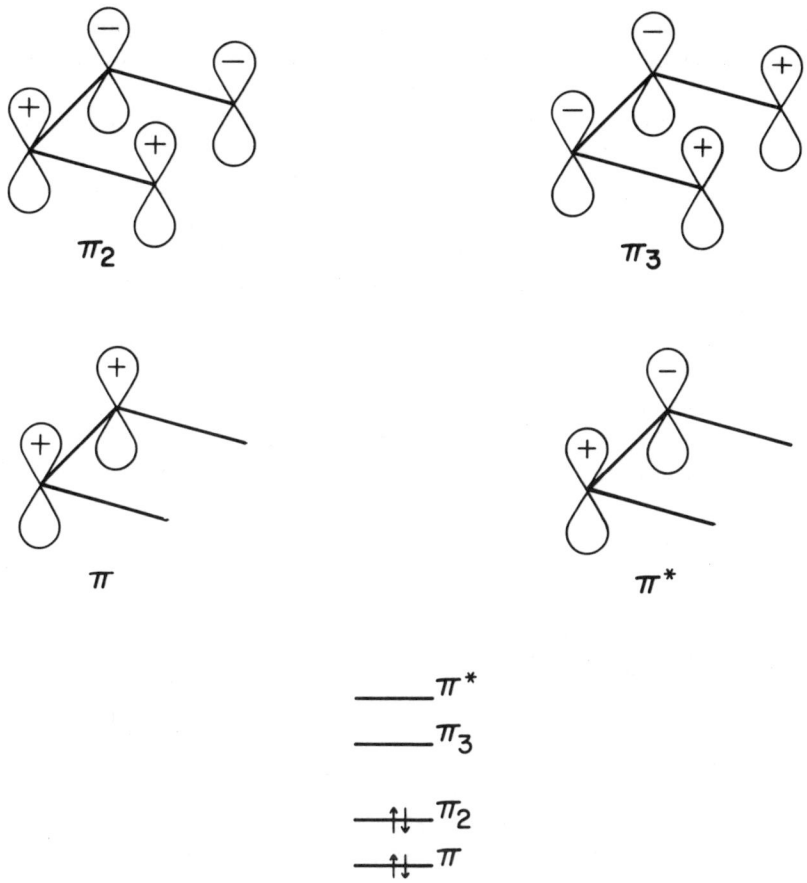

Figure 7-7
(a) Diene π_2 and π_3 orbitals. (b) Ene π and π^* orbitals. (c) Energy ordering of the orbitals in (a) and (b).

singlet configuration $\pi^2(\pi_2\pi_3)^1$ is considered. The orbitals π_2 and π_3 are now singly occupied molecular orbitals (SOMO) that can act either as *electron donors or acceptors* in the HOMO → LUMO excitation sense. Although *all* possible single excitations should be considered when determining how electron density can flow between two reactants to break old bonds and make new bonds, orbital energy differences (according to perturbation theory) influence the contributions to the overall electron density flow, and thus indicate that the $\pi \to \pi_2$ and $\pi_3 \to \pi^*$ excitations are probably more important. A $\pi_2 \to \pi$ excitation would put three electrons in the π orbital (which is not allowed by the Pauli principle), whereas the $\pi \to \pi_3$ or $\pi_2 \to \pi^*$ excitation would require higher energy. The $\pi_2 \to \pi_3$, $\pi_3 \to \pi_2$, and $\pi \to \pi^*$ orbital promotions do not cause charge to flow *between* reactants, but they are entirely intrafragment excitations. Notice that, analogous to the perturbation treatment of charge flow in the ground state, here these orbital promotions are analyzed with respect to their producing useful electron flow *starting* from an excited state. Examination of the energetically favored $\pi^2 \to \pi_2$ and $\pi_3 \to \pi^*$ orbital excitations shows that these orbital-promotion pairs do not produce a favorable overlap that allows *new product bonds* to form. Hence, the photochemical reaction should be forbidden. Certainly all of the orbital excitations have *some* influence on the charge flow that accompanies this reaction, but the most significant factor is whether there are low-energy single excitations that form new bonds as old bonds break.

The same conclusion would be reached if the *ene* had been excited rather than the diene. The orbital occupancy would then be given by $\pi_2^2(\pi\pi^*)^1$, and the relevant energetically favored excitations would be $\pi^* \to \pi_3^*$ and $\pi_2 \to \pi$, both of which lead to unfavorable overlap and, hence, a forbidden reaction—that is, no formation of new bonds.

7.4. Excited Reactants Can Correlate Directly with Ground-State Products

By promoting an electron from a doubly occupied orbital whose energy is *increasing* along the reaction coordinate to one whose energy *decreases*, an S_1 surface might be obtained that has no symmetry-imposed barriers. In such a case, the system moves from the S_1 surface to the S_0 surface of products by intersystem crossing at the point where the S_1 surface intersects S_0 (or the upper cone of S_0) in a manner favoring product formation.

However, cases are known in which internal conversion is not needed. Consider, for example, the photochemical abstraction of a hydrogen atom from an alkane by an excited carbonyl group

Treating the *active* orbitals as the CO π and π^*, the O *p*-like lone pair (n_O), and the H—C σ and σ^* orbitals, the orbital-correlation diagram shown in Figure 7-8 can be made, in which only the approximate symmetry plane of the H_2CO moiety is used to label the orbitals. The orbitals on the C and O of the product that lie perpendicular to the COH plane are labeled π_C and π_O. This symmetry label is only correct while the carbonyl group remains planar—

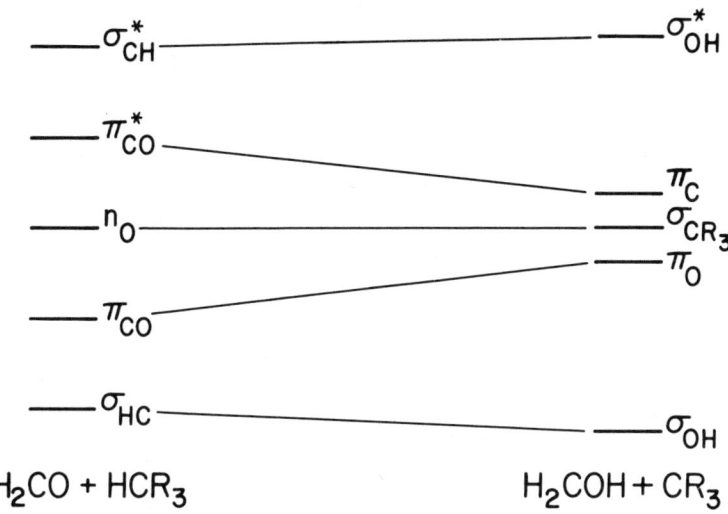

Figure 7-8
Orbital-correlation diagram for H-atom abstraction by carbonyl.

because when this radical becomes nonplanar, the orbitals no longer have pure π symmetry, π_O becomes a lone pair on oxygen, and π_C becomes a radical orbital on carbon having a mixed p and s character.

The orbital-correlation diagram in Figure 7-8 *seems* to indicate that the thermal reaction is allowed. However, notice that the $\sigma_{CH}^2 \pi_{CO}^2 n_O^2$ configuration correlates with the $\sigma_{OH}^2 \pi_O^2 \sigma_{CR_3}^2$ configuration of the products and that the product configuration also corresponds to the *ionic* products H_2COH^+ and C^-R_3. Clearly, these ionic species could not be the ground state of the products unless extreme solvation effects were present. (We shall not consider how solvation can affect the S_n and T_n surfaces; this is a separate, but very important, topic that is beyond the scope of the present work.) Hence, the orbital-correlation diagram does *not* predict that ground-state reactants can smoothly give rise to ground-state (radical) products.

The various low-energy configurations that can arise by occupying the reactant and product molecular orbitals in various ways are shown in the *singlet* configuration-correlation diagram shown in Figure 7-9. Notice that this diagram is different from those that have been encountered in that the singlet

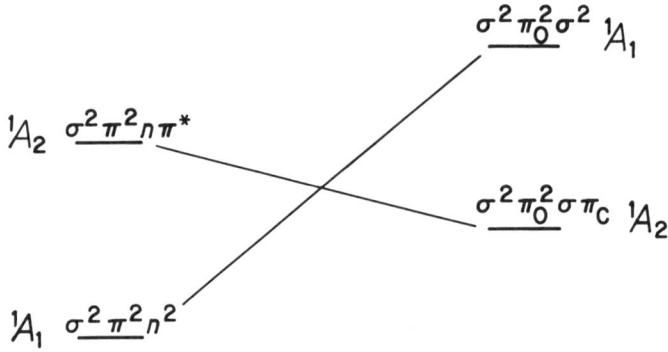

Figure 7-9
Configuration-correlation diagram for H-atom abstraction by carbonyl.

$(n\pi^*)^1$ excited configurations of the reactants *directly correlate* with the ground-state configuration of the product. This means that an efficient *direct* mechanism is available for bringing about the reaction

$$S_0 + h\nu \rightarrow S_1 \text{ (reactants)} \rightarrow S_0 \text{ (products)}.$$

No internal conversion or funneling is needed, and photochemical excitation of a $(n\pi^*)^1$ $^1A_2$ reactant can lead directly to *ground-state* radical products.

Had we included the triplet $(n\pi^*)^3$ configuration of the reactants on this configuration-correlation diagram, it would have correlated directly with the triplet $\sigma^2\pi^2\sigma\pi$ state of the product. Thus, triplet sensitized carbonyls should also abstract hydrogen atoms from alkanes as long as the triplet $(n\pi^*)^3$ state of the reactants is above the ground (singlet or triplet radical) state of the products (Turro, 1978).

7.5. Benzene Photochemistry

In this example, three rearrangements of benzene shown in Figure 7-10 are considered. The dewarbenzene and prismane are drawn both in their proper three-dimensional structures and symbolically as valence isomers of planar benzene to emphasize the bonding relationships among these species. We begin the analysis by proposing reaction coordinates for each of the above reactions that preserve C_{2v} symmetry and by labeling and ordering the active orbitals of the three species.

In order of increasing energy from left to right, for benzene (see Cotton, 1963), the π molecular orbitals are

EXAMPLES OF PHOTOCHEMICAL REACTIONS   107

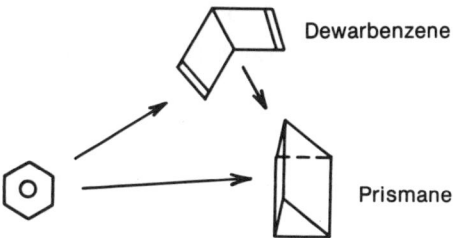

Figure 7-10
Valence isomers of benzene.

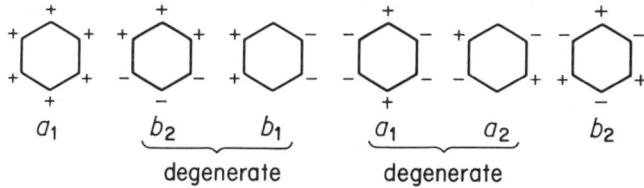

in which the + and − signs label the relative signs of the (p_π) atomic orbitals. The σ and π orbitals of dewarbenzene are

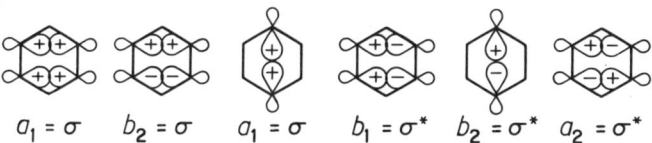

and the three σ orbitals of prismane are

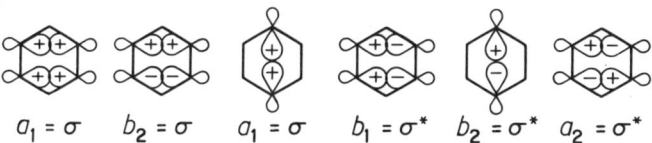

These sets of orbitals lead to the orbital-correlation diagram shown in Figure 7-11. The diagram is *not* drawn to be quantitatively accurate but indicates that σ bonds are stronger than π bonds; furthermore, the orbital-correlation diagram contains no information about the *strain energy* of 1,4-dewarbenzene or prismane. Construction of a configuration-correlation diagram appropriate for the ground and low-lying *singlet* excited states of these three valence isomers is begun by listing the configurations expected to be most important, indicating the *essential configurations* for each of the three species and the correlations among these and other configurations (Table 7-1).

TABLE 7-1
Correlation of Essential Configurations for Benzene Valence Isomers

Configuration	Dominant species	Correlates with
$a_1^2 b_2^2 a_1^2$	Prismane	$\pi^4 \pi^{*2}$ benzene and $\sigma^2 \pi^2 \pi^{*2}$ dewarbenzene
$a_1^2 b_2^2 a_1 b_1$	Prismane*	$\pi^5 \pi^*$ benzene and $\sigma^2 \pi^2 \pi^{*2}$ dewarbenzene
$a_1^2 b_2^2 b_1^2$	Benzene	$\sigma^4 \sigma^{*2}$ prismane and $\sigma^2 \pi^2 \pi^{*2}$ dewarbenzene
$a_1^2 b_2^2 b_1 a_1$	Benzene*	$\sigma^5 \sigma^*$ prismane and $\sigma^2 \pi^2 \pi^{*2}$ dewarbenzene
$a_1^2 a_1^2 b_1^2$	Dewarbenzene	$\sigma^4 \sigma^{*2}$ prismane and $\pi^4 \pi^{*2}$ benzene
$a_1^2 a_1^2 b_1 b_2$	Dewarbenzene*	$\sigma^5 \sigma^*$ prismane and $\pi^4 \pi^{*2}$ benzene
$a_1^2 a_1 b_1^2 b_2$	Dewarbenzene*	$\sigma^4 \sigma^{*2}$ prismane and $\pi^5 \pi^*$ benzene

Note: The * indicates an electronically excited molecule or orbital.

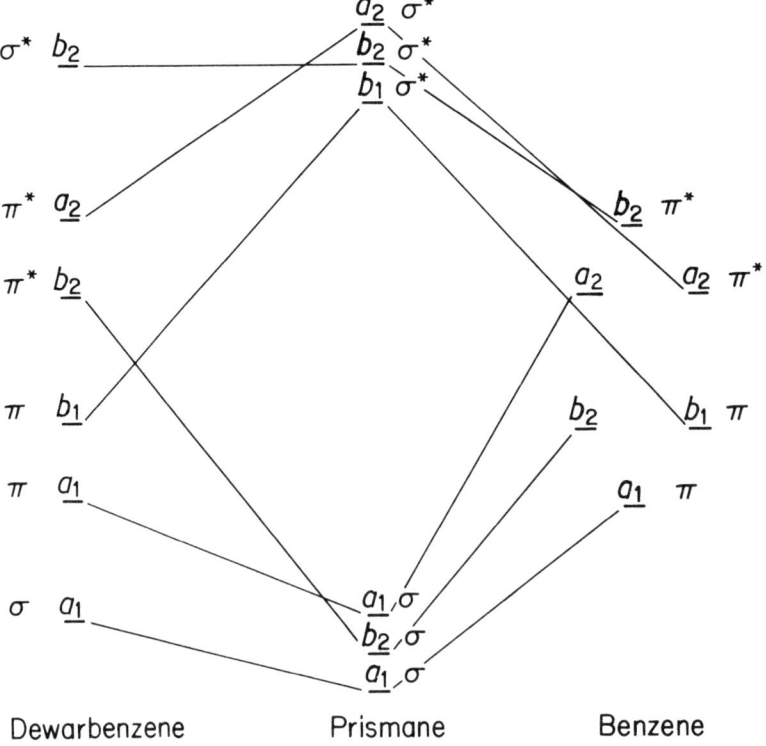

Figure 7-11
Orbital-correlation diagram for rearrangements of benzene.

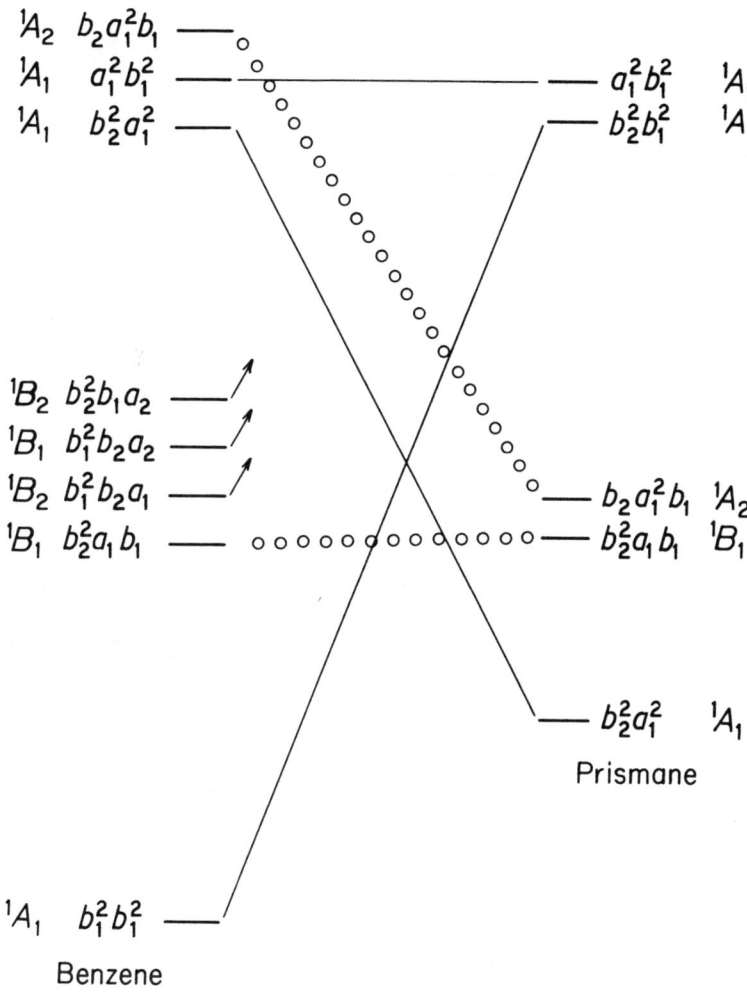

Figure 7-12
Configuration-correlation diagram for benzene and prismane.

This information is then used to construct configuration-correlation diagrams for the three reactions (Figures 7-12, 7-13, and 7-14). Since the lowest energy orbitals of all three systems have a_1 symmetry and are doubly occupied, the two electrons in these orbitals are neglected in constructing the configuration-correlation diagrams. In each diagram the configurations that

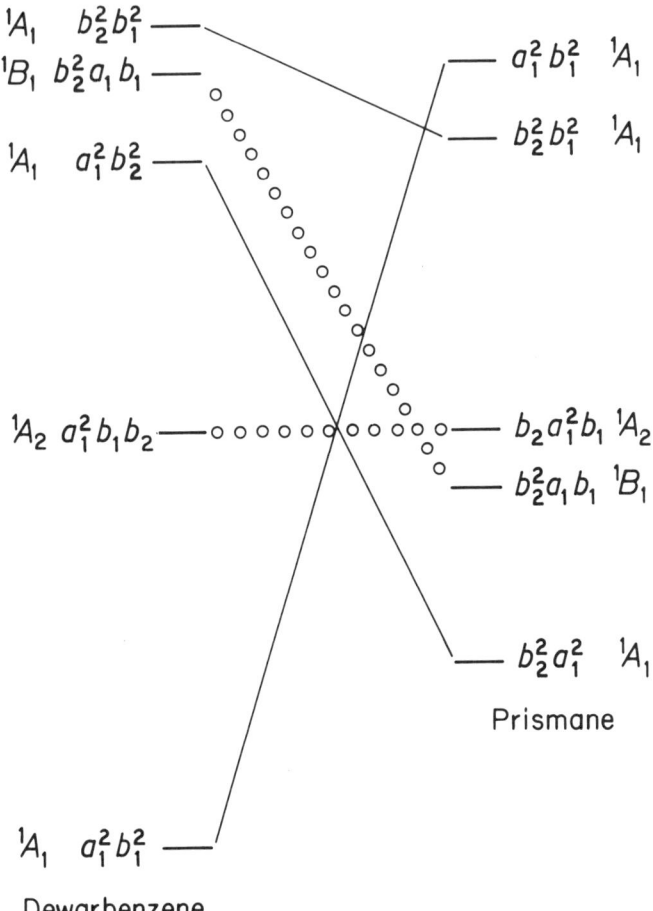

Figure 7-13
Configuration-correlation diagram for dewarbenzene and prismane.

contribute to S_0 of reactants or products and those that give rise to low-lying singly excited states are displayed. Arrows directed steeply upwards indicate that the particular configuration correlates with a doubly or more highly excited configuration on the other side of the diagram.

These configuration-correlation diagrams indicate that the $^1B_1$ excited state of benzene would *not* lead to dewarbenzene for photon energies near the $^1B_1$ absorption threshold; although the process is not symmetry-forbidden, it is a very uphill process that leads to the doubly excited $b_2^2 a_1 b_1$ state of dewarbenzene. Along the uphill movement, the $^1B_1$ surface crosses the $^1B_2$

EXAMPLES OF PHOTOCHEMICAL REACTIONS 111

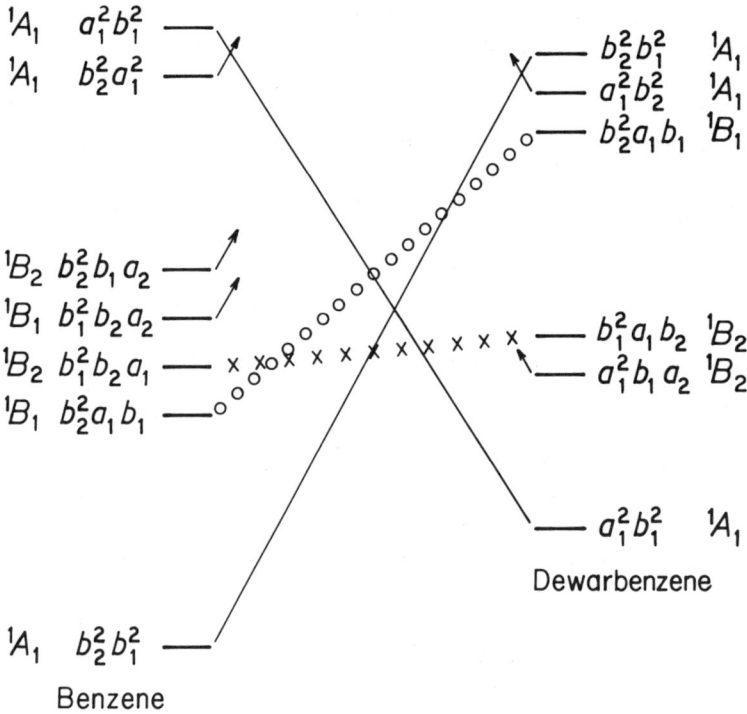

Figure 7-14
Configuration-correlation diagram for benzene and dewarbenzene.

surface. If the photon energy were sufficient to place the system high enough on the $^1B_1$ surface, the system could hop—by a vibration of $b_1 \times b_2 = a_2$ symmetry—to the $^1B_2$ surface. The system might then move on this $^1B_2$ surface, eventually crossing again to the $^1A_1$ ground-state surface. Since the $^1B_2$-to-$^1A_1$ surface crossing occurs on the reactant side of the $^1A_1$ surface, ground-state benzene should be the main product.

Alternatively, this same benzene-excitation process *could* yield either ground-state benzene or prismane because the $^1B_1$ surface crosses the $^1A_1$ surface along the reaction coordinate connecting benzene and prismane. On the other hand, one of the benzene $^1B_2$ configurations correlates directly with a low-lying $^1B_2$ configuration of dewarbenzene and should thus give an allowed reaction that might yield *some* S_0 dewarbenzene (but no prismane), the relative amounts depending on the exact values of quantum yields. If the $^1A_1$ S_0 surface is intersected by the $^1B_2$ surface on the benzene side, then intersystem crossing will most likely lead to ground-state benzene. On the other hand, if the top of the S_0 surface lies below the $^1B_2$ surface, formation of ground-state dewarbenzene is equally likely.

Excitation to the lowest $^1A_2$ state of dewarbenzene should yield an appreciable amount of ground-state prismane, since the lowest $^1A_2$ configuration of each species correlates directly and crosses the funnel region of the $^1A_1$ ground state. At higher photon energies, excitation of a $^1B_2$ state of dewarbenzene can give rise to formation of ground-state benzene; this is the reverse of the reaction just discussed.

The number of events that *might* occur when a photon is absorbed is quite large even for a system having few low-energy excited states. A small number of occupied orbitals out of which an electron can be excited and a small number of low-energy virtual orbitals can give rise to a large number of singly excited states. Moreover, a number of geometrical distortions (that is, proposed reaction coordinates) may have to be considered in following reactant states through to various product states (as in the case just discussed). The crossings of the excited potential-energy curves having low energy among one another *and* with the ground state (S_0) surface along the possible reaction coordinates determine the quantum yields of the numerous available reactive, radiative, and radiationless pathways. Although the symmetry and nodal-pattern tools do not allow a *quantitative* prediction of the yields of the competing events, they allow one to guess the events that are likely and those that are not likely because of symmetry-imposed barriers.

7.6. C + H$_2$ → CH$_2$

In this example, the reactions $^1A_1\text{CH}_2 \to \text{H}_2 + \text{C}$ and $^3B_1\text{CH}_2 \to \text{H}_2 + \text{C}$ are investigated in an *assumed* C_{2v} reaction pathway. To form the appropriate correlation diagrams the following information is needed (all energies are in kcal/mole):

$\text{C}(^3P) \to \text{C}(^1D)$, $\Delta E = 29.2$

$\text{C}(^1D) \to (^1S)$, $\Delta E = 32.7$

$\text{C}(^3P) + \text{H}_2 \to \text{CH}_2(^3B_1)$, $\Delta E = -78.8$

$\text{C}(^1D) + \text{H}_2 \to \text{CH}_2(^1A_1)$, $\Delta E = -97.0$.

Using the coordinate system shown in Figure 7-15, the hydrogen σ_g and σ_u orbitals and the carbon 2s, $2p_x$, $2p_y$, and $2p_z$ orbitals are labeled as either a_1, b_1 or b_2, as are the σ, σ, σ^*, σ^*, n, and p_π orbitals of CH$_2$. For the reactants,

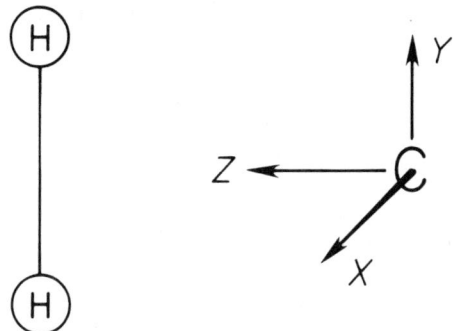

Figure 7-15
Coordinate system for discussion the reaction C + H$_2$ → CH$_2$.

the σ_g and σ_u orbitals have a_1 and b_2 symmetry, respectively. The nitrogen orbitals and their symmetries are $2s$ (a_1), $2p_z$ (a_1), $2p_y$ (b_2), and $2p_x$ (b_1). For CH$_2$ the symmetric combinations of the two CH σ and σ^* bonds have a_1 symmetry, the antisymmetric σ and σ^* combinations have b_2 symmetry, the nonbonding (n) orbital has a_1 symmetry, and the $p_\pi(x)$ orbital has b_1 symmetry.

An orbital-correlation diagram for the CH$_2$ → C + H$_2$ reactions can be drawn in which the orbitals are ordered by their relative energies. The same orbital-correlation diagram applies to both reactions; only in the configuration- and state-correlation diagrams does one distinguish between the triplet and singlet species. The orbital-correlation diagram is constructed by connecting orbitals having the same symmetry, as shown in Figure 7-16. To proceed, ^3P, ^1D, and ^1S wavefunctions of the carbon atoms must be symmetry-analyzed. We write these wavefunctions—first in terms of $2p_m$ where $m = 1, 0, -1$ orbitals and then in terms of $2p_{x,y,z}$ orbitals. These wavefunctions are the following:

1. Three Slater-determinant wavefunctions belonging to the ^3P state, each of which has an M_s value of 1. (Any value of M_s—1, 0, or −1— could be chosen because the reaction of ^3P C to produce $^3B_1$ CH$_2$ is independent of M_s.)
2. Five ^1D Slater-determinant wavefunctions
3. One ^1S Slater-determinant wavefunction

Then, the configuration-correlation diagrams for the above singlet and triplet reactions can be constructed.

The $M_s = 1$ functions all have two unpaired α electrons. If the $1s^2$ and $2s^2$ spin orbitals, which are common to the first four columns of each Slater determinant are ignored, for ^3P (M_L, M_s)

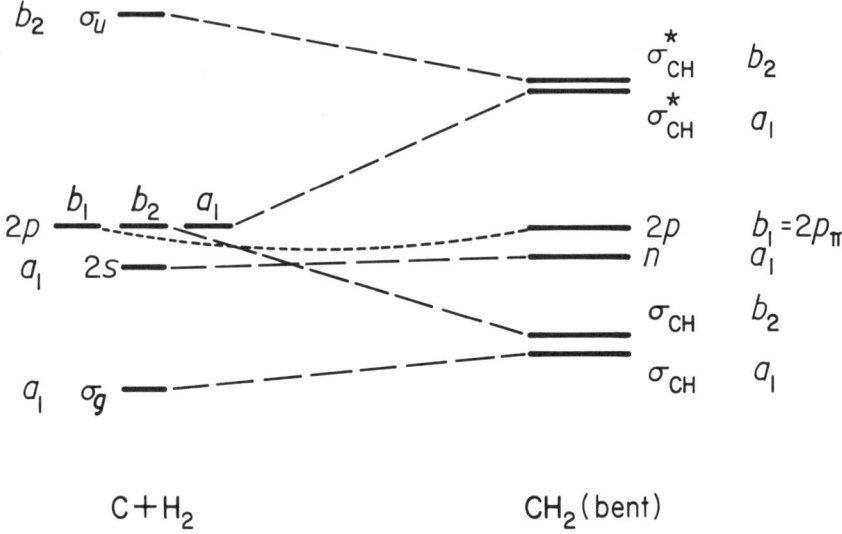

Figure 7-16
Orbital-correlation diagram for the reaction C + H$_2$ → CH$_2$.

$$^3P(1,1) = |2p\alpha_{+1}2p_0\alpha|$$

and

$$^3P(0,1) = |2p_{+1}\alpha 2p_{-1}\alpha|$$

$$^3P(-1,1) = |2p_0\alpha 2p_{-1}\alpha|.$$

The five ^1D determinantal functions are obtained as follows. The $^1D(2,0)$ and $^1D(-2,0)$ functions are the only determinants arising from p^2 that have $M_L = \pm 2$ and are

$$^1D(2,0) = |2p_{+1}\alpha 2p_{+1}\beta| \text{ and } ^1D(2,0) = |2p_{-1}\alpha 2p_{-1}\beta|.$$

The other $^1D(M_L,0)$ $M_L = 1,0,-1$ functions are obtained by applying the L_- lowering operator to $^1D(2,0)$ or the L_+ raising operator to $^1D(-2,0)$. In so doing, we use the fact that L_+ acting on an eigenfunction of L^2 and L_z yields a multiple of the eigenfunction having one higher or lower L_z eigenvalue. Thus, $L_-{}^1D(2,0) \sim {}^1D(1,0)$, and $L_+{}^1D(-2,0) \sim {}^1D(-1,0)$.

The operators $L_\pm$ are most conveniently expressed as sums of orbital-level operators $L_\pm = \Sigma_i l_\pm(i)$. The effect of $l_\pm$ on any of the closed-shell $(1s^2 2s^2)$ components of the above Slater determinants need not be considered; only the

$2p^2$ aspects are treated. Since $l_\pm$ operating as an orbital having (l,m) eigenvalues gives $\hbar\sqrt{l(l+1)-m(m\pm 1)}$ times the orbital with eigenvalues $(l, m\pm 1)$,

$$L_-^1D(2,0) = [l_-(5)+l_-(6)]|2p_{+1}\alpha 2p_{+1}\beta|$$
$$= \hbar\sqrt{2}[|2p_0\alpha 2p_{+1}\beta| + |2p_{+1}\alpha 2p_0\beta|] \sim {}^1D(1,0)$$

and

$$L_+{}^1D(-2,0) = [l_+(5)+l_+(6)]|2p_{-1}\alpha 2p_{-1}\beta|$$
$$= \hbar\sqrt{2}[|2p_0\alpha 2p_{-1}\beta| + |2p_{-1}\alpha 2p_0\beta|] \sim {}^1D(-1,0)$$

To obtain ${}^1D(0,0)$ we can apply L_- to ${}^1D(1,0)$ or L_+ to ${}^1D(-1,0)$. For example,

$$L_-{}^1D(1,0) = \hbar\sqrt{2}[|2p_{-1}\alpha 2p_{+1}\beta| + 2|2p_0\alpha 2p_0\beta|$$
$$+ |2p_{+1}\alpha 2p_{-1}\beta|]\hbar\sqrt{2} \sim {}^1D(0,0).$$

These combinations of Slater determinants are not normalized; to normalize them is straightforward, for example,

$${}^1D(0,0) = \frac{1}{\sqrt{6}}[|2p_{+1}\alpha 2p_{-1}\beta| + |2p_{-1}\alpha 2p_{+1}\beta| + 2|2p_0\alpha 2p_0\beta|].$$

The one ${}^1S(0,0)$ Slater-determinant function can be obtained as the remaining combination of the determinants having $M_L = 0$, $M_s = 0$ that is orthogonal to ${}^1D(0,0)$ and ${}^3P(0,0)$. Recall that ${}^1D(0,0)$ is given above. ${}^3P(0,0)$ can be obtained from ${}^3P(0,1)$ by applying S_-:

$${}^3P(0,0) \sim S_-{}^3P(0,1) = (S_-(5)+S_-(6))|2p_{+1}\alpha 2p_{-1}\alpha|$$
$$= \hbar\sqrt{1}[|2p_{+1}\beta 2p_{-1}\alpha| + |2p_{+1}\alpha 2p_{-1}\beta|].$$

Clearly ${}^1D(0,0)$ has the form $2z+x+y$, whereas ${}^3P(0,0)$ contains $x-y$; hence, ${}^1S(0,0)$ must have the form $z-x-y$, or

$${}^1S(0,0) = \frac{1}{\sqrt{3}}[|2p_0\alpha 2p_0\beta| - |2p_{+1}\alpha 2p_{-1}\beta| - |2p_{-1}\alpha 2p_{+1}\beta|].$$

To express the ${}^1D$, ${}^3P(M_L,1)$, and ${}^1S$ wavefunctions in terms of $2p_{x,y,z}$ orbitals, we must write out each Slater determinant and substitute $2p_z$ for $2p_0$

and $(2p_x \pm i2p_y)2^{-1/2}$ for $2p_{\pm 1}$; this will generate Slater determinants including $2p_{x,y,z}$ orbitals. It is important to understand the reason for bringing about this transformation from $m_l = 1, 0, -1$ to x,y,z space. The $2p_m$ orbitals and their determinental wavefunctions are appropriate for the spherically symmetrical carbon atom in which $L_z = \Sigma_i l_z(i)$ commutes with the electronic Hamiltonian. However, in the presence of the H_2 molecule in C_{2v} symmetry, L_z no longer commutes with the electronic Hamiltonian, though the operations of the C_{2v} point group (E, σ_v, $\sigma_{v'}$, C_2) do. Because the $2p_{x,y,z}$ orbitals are symmetry-adapted with respect to C_{2v} symmetry, these orbitals must be used in the wavefunctions.

The transformations of the three $^3P(M_L,1)$, five $^1D(M_L,0)$, and one $^1S(0,0)$ wavefunctions to x,y,z-space are the following. The transformations of the P wavefunctions are

$$^3P(1,1) = |2p_{+1}\alpha 2p_0\alpha| = 2^{-1/2}[|2p_x\alpha 2p_z\alpha| + i|2p_y\alpha 2p_z\alpha|]$$

$$^3P(-1,1) = 2^{-1/2}[|2p_x\alpha 2p_z\alpha| - i|2p_y\alpha 2p_z\alpha|]$$

$$^3P(0,1) = 2^{-1}[|2p_x\alpha 2p_x\alpha| + |2p_y\alpha 2p_y\alpha| + i|2p_y\alpha 2p_x\alpha| - i|2p_x\alpha 2p_y\alpha|]$$

$$= i|2p_y\alpha 2p_x\alpha|.$$

For $^3P(0,1)$ we used the facts that Slater determinants are antisymmetric

$$|2p_x\alpha 2p_y\alpha| = -|2p_y\alpha 2p_x\alpha|$$

and that they obey the Pauli principle

$$|2p_x\alpha 2p_x\alpha| = 0$$

The *three* $^3P(M_L,1)$ functions are degenerate when the H_2 molecule is not present, so any combinations of these three functions would also be degenerate. In particular,

$$[^3P(1,1) + {}^3P(-1,1)]2^{-1/2} = |2p_x\alpha 2p_z\alpha| \equiv {}^3P(xz,1)$$

$$\frac{1}{i}[^3P(1,1) - {}^3P(-1,1)]2^{-1/2} = |2p_y\alpha 2p_z\alpha| \equiv {}^3P(yz,1)$$

and

$$\frac{1}{i}{}^3P(0,1) = |2p_y\alpha 2p_x\alpha| \equiv {}^3P(yx,1)$$

are all degenerate. However, the three new functions (yz, xz, and yx) are more useful because they are symmetry-adapted for C_{2v} symmetry: $^3P(xz,1)$ $^3P(yz,1)$ and $^3P(yx,1)$ have B_1, B_2 and A_2 symmetry, respectively. These symmetries are obtained as the direct products of the symmetries of the orbitals. For example, $^3P(yx,1)$ has A_2 symmetry because the direct product of $b_2(y)$ and $b_1(x)$ has A_2 symmetry. Notice that it is *not* correct to conclude that the ^3P state functions would span the same symmetry space as three $2p$ orbitals do (p_x has b_1, p_y has b_2 and p_z has a_1 symmetry). A second reason that the xz, yz, and yx determinants are more useful for the C_{2v} case is that they correspond to a single orbital occupancy from which configuration-correlation diagrams are easily generated.

Let us now consider the five degenerate $^1D(M_L,0)$ wavefunctions.

$$^1D(2,0) = |2p_{+1}\alpha 2p_{+1}\beta| = \frac{i}{2}[|2p_x\alpha 2p_y\beta| + |2p_y\alpha 2p_x\beta|]$$

$$+ \frac{1}{2}[|2p_x\alpha 2p_x\beta| - |2p_y\alpha 2p_y\beta|]$$

$$^1D(-2,0) = |2p_{-1}\alpha 2p_{-1}\beta| = \frac{1}{2}[|2p_x\alpha 2p_x\beta| - |2p_y\alpha 2p_y\beta|]$$

$$- \frac{i}{2}[|2p_x\alpha 2p_y\beta| + |2p_y\alpha 2p_x\beta|].$$

$$^1D(1,0) = 2^{-1/2}[|2p_0\alpha 2p_{+1}\beta| + |2p_{+1}\alpha 2p_0\beta|]$$

$$= \frac{1}{2}[|2p_z\alpha 2p_x\beta| + |2p_x\alpha 2p_z\beta| + i|2p_z\alpha 2p_y\beta| + i|2p_y\alpha 2p_z\beta|]$$

$$^1D(-1,0) = \frac{1}{2}[|2p_z\alpha 2p_x\beta| + |2p_x\alpha 2p_z\beta| - i|2p_z\alpha 2p_y\beta| - i|2p_y\alpha 2p_z\beta|]$$

$$^1D(0,0) = 6^{-1/2}[2|2p_0\alpha 2p_0\beta| + |2p_{+1}\alpha 2p_{-1}\beta| + |2p_{-1}\alpha 2p_{+1}\beta|]$$

$$= 6^{-1/2}[2|2p_z\alpha 2p_z\beta| + |2p_x\alpha 2p_x\beta| + |2p_y\alpha 2p_y\beta|].$$

Since degenerate functions can be combined without affecting their degeneracy, $^1D(\pm 2,0)$ can be combined to yield functions having symmetries $^1D(xx,0) - ^1D(yy,0)$ and $^1D(xy,0)$. Likewise, $^1D(\pm 1,0)$ can be combined to yield $^1D(xz,0)$ and $D(yz,0)$ symmetry functions. These new combinations are useful because they are symmetry-adapted: For example, the symmetries of $^1D(xx,0) - ^1D(yy,0)$, $^1D(xy,0)$, $^1D(xz,0)$, and $^1D(yz,0)$ are A_1, A_2, B_1 and B_2, respectively; the $^1D(0,0)$ function has A_1 symmetry.

118 CHAPTER 7

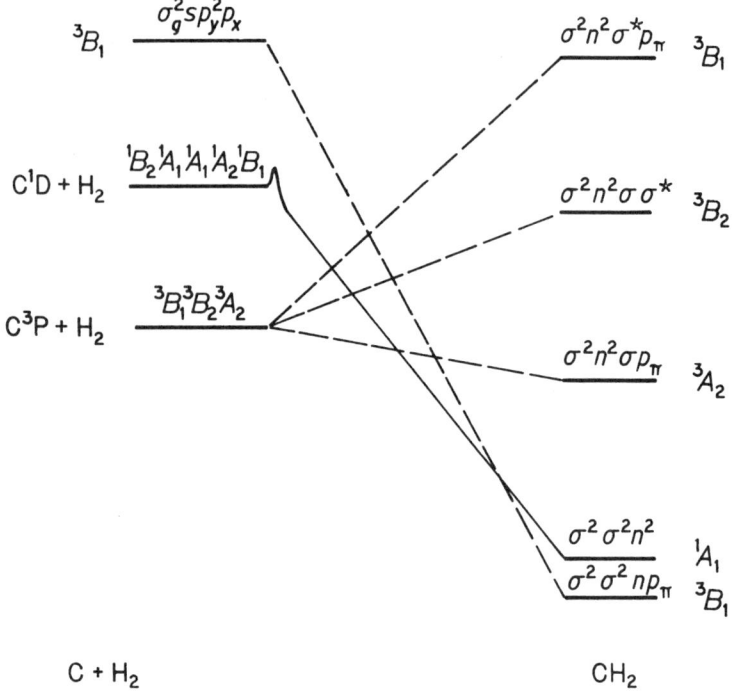

Figure 7-17
Configuration-correlation diagram for singlet and triplet.

The $^1S(0,0)$ wavefunction has the property that

$$^1S(0,0) = 3^{-1/2}[|2p_0\alpha 2p_0\beta| - |2p_{+1}\alpha 2p_{-1}\beta| - |2p_{-1}\alpha 2p_{+1}\beta|]$$

$$= 3^{-1/2}[|2p_z\alpha 2p_z\beta| - |2p_x\alpha 2p_x\beta| - |2p_y\alpha 2p_y\beta|],$$

which has A_1 symmetry in C_{2v} since each of its components has A_1 symmetry—that is, $a_1 \times a_1 = A_1$, $b_1 \times b_1 = A_1$, $b_2 \times b_2 = A_1$.

The configuration-correlation diagram, together with the relevant avoided crossings, is shown in Figure 7-17. In constructing this diagram, the relative energy orderings of various configurations must be kept in mind. For example, the $\sigma^2 n^2 p_x \sigma$ configuration is placed lower then the $\sigma^2 n^2 \sigma\sigma^*$ one, which, in turn, is lower than $\sigma^2 n^2 p_x \sigma^*$.

Let us examine how the configurations have been correlated. The $^3B_1$ $\sigma^2\sigma^2 np_x$ configuration of CH$_2$ correlates with the $\sigma_g^2 p_y^2 2sp_x$ configuration of C + H$_2$ (see the orbital-configuration diagram in Figure 7-16). The latter configuration does not even arise from H$_2$ + C($1s^2 2s^2 2p^2$) but from

$H_2 + C(1s^22s2p^3)$ and, as such, lies considerably above the ^1S state of $C(1s^22s^22p^2)$.

The $^1A_1$ $\sigma^2\sigma^2n^2$ configuration of CH_2 correlates with $\sigma_g^2 2s^2 2p_y^2$ (see Figure 7-16). This latter configuration is *not* purely ^1D or ^1S. It is a combination of ^1D and ^1S functions—in particular

$$|\sigma_g^2 2s^2 2p_y^2| = 6^{-1}[6^{1/2}\ ^1D(0,0) - 3^{1/2} 2^1S(0,0)] - 2^{-1}[^1D(2,0) + {}^1D(-2,0)],$$

which is 2/3 ^1D and 1/3 ^1S in character. Hence, the configuration-correlation diagram must be drawn with a barrier near the ^1D asymptote to represent the fact that $^1A_1$ CH_2 correlates with a (2/3:1/3) *mixture* of ^1D and ^1S C (plus H_2), which will eventually mix (with $2p_x^2$ and $2p_z^2$) to yield the ^1D and ^1S states.

The $^3A_2$ $\sigma^2 n^2 \sigma p_x$ and $^3B_2$ $\sigma^2 n^2 \sigma\sigma^*$ configurations of CH_2 correlate with $\sigma_g^2 2s^2 p_x p_y$ and $\sigma_g^2 2s^2 p_y p_z$ configurations of $C + H_2$. The latter two triplet configurations are members of the three degenerate $^3P(xy,1)$, $^3P(yz,1)$ and $^3P(xz,1)$ functions. The third member of this family—the $^3P(xz,1)$ configuration $\sigma_g^2 2s^2 p_x p_z$—which has $^3B_1$ symmetry—correlates with the $\sigma^2 n^2 p_x \sigma^*$ configuration of CH_2.

In like fashion, all five of the ^1D and the ^1S states of $C + H_2$ can be correlated with those of CH_2. However, since we are considering only the *lowest* triplet and singlet states, this correlation is unnecessary. All that needs to be done is to seek low-energy configurations that have $^1A_1$ or $^3B_1$ symmetry in the C_{2v} point group.

We now examine whether the reactions $C(^3P) + H_2 \rightarrow CH_2$ and $C(^1D) + H_2 \rightarrow CH_2$ have large activation barriers and determine the states of CH_2 that are produced in these reaction.

The configuration-correlation diagram in Figure 7-17 clearly illustrates that the $^3B_1$ reaction—$C + H_2 \rightarrow CH_2$—should have a symmetry-imposed barrier. The transition state along this reaction path should lie closer to the $C + H_2$ reactants than to the CH_2 products because the forward reaction is exothermic.

The $^1A_1$ reaction $C(^1D) + H_2 \rightarrow CH_2$ also has a symmetry barrier in the configuration-correlation diagram, but this barrier is artifical. Recall that the $^1A_1\sigma^2\sigma^2n^2$ CH_2 configuration correlates with a 2/3:1/3 mixture of ^1D and ^1S configurations. Hence, as $CH_2(^1A_1)$ is pulled apart along the assumed C_{2v} reaction path, the electronic wavefunction must mix the $^1D(0,0)$, $^1D(+2,0) + {}^1D(-2,0)$, and the $^1S(0,0)$ configurations. However, as the distance between the C and H_2 species becomes so large that they no longer interact, the wavefunction smoothly evolves to have only (^1D) $C + H_2$ symmetry.

Another aspect of the configuration correlation diagram shown in Figure 7-17 is of interest. The lowest $^3B_1$ surface, which has a substantial barrier, is crossed by the $^3A_2$ and $^3B_2$ surfaces, so we may ask whether pseudo-Jahn-Teller effects occur. The $^3B_1{}^3B_2$ crossing would require a $B_1 \times B_2 = A_2$

vibrational distortion to give rise to mixing. The CH_2 molecule has no vibration with this symmetry. In constrast, the $^3B_1{}^3A_2$ crossing could give rise to a pseudo-Jahn-Teller effect through a distortion having $A_2 \times B_1 = B_2$ symmetry. The asymmetric stretch vibration of CH_2 has b_2 symmetry; this means that as the CH_2 is pulled apart, any asymmetric stretch motion could cause a transition from the $^3B_1$ surface to the $^3A_2$ surface, after which further dissociation could occur on $^3A_2$ to give rise to $C(^3P) + H_2$. The result of such a surface transition would be a lowering of the activation energy of the dissociation reaction. The $^3B_1 \rightarrow {}^3A_2$ transition need not take place in every collision. Those collisions in which the molecule ends up on the $^3A_2$ surface will experience a lower barrier.

Such pseudo-Jahn-Teller effects can also affect the reaction $C(^3P) + H_2 \rightarrow CH_2$. Those C atoms whose orbital occupancy is $p_x p_y$ ($^3A_2$) can follow the $^3A_2$ surface, which has no barrier, until the $^3A_2{}^3B_1$ crossing. At the crossing the asymmetric distortion can permit the system to move to the $^3B_1$ surface and thereby form ground-state CH_2 products. Those $C(^3P)$ atoms whose orbital occupancies are $p_x p_z(^3B_1)$ or $p_y p_z(^3B_2)$ will encounter barriers as the H_2 approaches. The $^3B_2$ surface appears to have the smallest barrier (activation energy) but, as mentioned above, the $^3B_2{}^3B_1$ crossing cannot give rise to a surface transition because CH_2 has no $B_2 \times B_1 = A_2$ vibration, and thus, $^3B_2$ collisions are ineffective. $^3B_1$ collisions can proceed directly (that is, with no pseudo-Jahn-Teller effects required) through the barrier on this surface to give ground-state CH_2 products.

Problems

1. You are studying the photochemical reaction in which 1,4-dewarnaphthalene rearranges in a disrotatory ring opening to yield naphthalene. The relevant energy change is $\Delta E = -48$ kcal/mol.

   $\Delta E = -48$ kcal/mole

   a. Using the one symmetry plane that is conserved in the reaction, draw and label as even (a') or odd (a'') all of the active orbitals of the benzene moiety and of the

   moiety. For example,

EXAMPLES OF PHOTOCHEMICAL REACTIONS   121

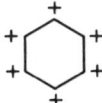

is a'.

b. The energy ordering of the orbitals in the dewar structure is a', a', a', a'', a', a'', a', a'', a'', a''. Describe the physical characteristics (e.g., σ or π, bonding or antibonding) of each of these orbitals in terms of the benzene and

moiety orbitals.

c. The energy ordering of the π orbitals in naphthalene is a', a', a'', a', a'', a', a'', a', a'', a''. Draw an orbital-correlation diagram for the reaction, labeling each orbital as a' or a'' *and* state the nature (σ, π, σ^*, or π^*) of each orbital.

d. The low-energy excited states of 1,4-dewarnaphthalene lie at 125 kcal/mol, 150 kcal/mol, and 180 kcal/mol and are triplet, singlet, and singlet, respectively. Their spatial symmetries are A'', A', and A'', respectively. Assign configurations to each of these three excited states and state the configurations of naphthalene with which they correlate. In all cases use the ground state of naphthalene as the reference point of energy.

e. The low-energy excited states of naphthalene lie 60 kcal/mol, 90 kcal/mol, and 100 kcal/mol above the ground state and have $^3A''$, $^1A'$, and $^1A''$ symmetries, respectively. Assign configurations to each of these three states and state the configurations of 1,4-dewarnaphthalene with which they correlate.

f. Draw a quantitatively correct configuration-correlation diagram using all of the above data. Give spin and space (A' or A'') labels to all configurations. Show how the configurations will mix to give rise to states. You may assume the doubly excited configurations lie 180 kcal/mol or more above their ground state configurations.

Now, based upon your state-correlation diagram, answer the following:

g. When light of 2850 Å is used to excite the dewarnaphthalene, why does one obtain primary fluorescence of the dewarnaphthalene?

h. What *other* fluorescence would you expect to see if the wavelength of the exciting light decreases to 2550 Å? Why? What does the observation of fluorescence at 3195 Å tell you about how the internal energy has been distributed within the excited dewarnaphthalene molecule?

i. At a much longer time after creating the initial excited state of the dewarnaphthalene, why does one see phosphorescence *only* from naphthalene?

2. The photochemistry of formaldehyde has received much attention recently. It is a "testing molecule" for models of energy-sharing, photodissociation, and internal

conversion. Let us try to understand some of the interesting features of this small molecule.

a. Draw an orbital-correlation diagram for the C_{2v} decomposition $H_2CO \rightarrow H_2 + CO$, labeling the orbitals according to their symmetry under the two reflection planes. Repeat this process for the $H_2CO \rightarrow H + HCO$ reaction, assuming the reaction to take place in a manner that preserves one symmetry plane. Include only the active orbitals in these diagrams.

The following facts are available: (1) The lowest $n\pi^*$ triplet and singlet excited states of H_2CO lie 25,200 cm^{-1} and 28,200 cm^{-1} above the ground state. (2) The CH bond energy in H_2CO is 88 kcal/mol. (3) $H_2CO \rightarrow H_2 + CO$ is exothermic by 11 kcal/mol. (4) H_2CO (^1A) $\rightarrow$ H(^2S) + HCO (linear $^2\pi$) is endothermic by 114 kcal/mol. (5) The lowest $n\pi^*$ singlet excited state of CO lies 65,500 cm^{-1} above its ground state.

b. Draw configuration-correlation diagrams for the C_{2v} $H_2CO \rightarrow H_2 + CO$ reaction and the C_s $H_2CO \rightarrow H + HCO$ (bent) reaction. Label the configurations according to symmetry and indicate how the configurations combine to give rise to states.

c. It is known that excitation of the singlet $n\pi^*$ state of H_2CO with light between 28,200 and 30,600 cm^{-1} leads to internal conversion, fluorescence, and formation of ground state $H_2 + CO$. By examining the C_{2v} correlation diagram, explain how $H_2 + CO$ could be formed. In particular, what kind of molecular deformation could be involved to allow (in a symmetry sense) the formation of ground-state $H_2 + CO$? Near 28,200 cm^{-1}, H_2CO^* undergoes fluorescence and internal conversion to ground-state H_2CO in a ratio fo 1:20. In contrast, D_2CO (D = deuterium) undergoes mostly fluorescence and very little internal conversion. Explain this difference between H_2CO and D_2CO. (An isotope effect is not a sufficient explanation).

d. As the energy of the exciting light reaches 30,600 cm^{-1}, formation of H + HCO becomes possible. On symmetry grounds, what kind of molecular deformation could give rise to these radical products? Be sure to explain the fact that the resulting HCO is bent.

e. Describe the mechanism by which triplet $n\pi^*H_2CO$ (formed by triplet sensitization) quickly gives rise to ground-state singlet H_2CO.

Appendix A

Overview of *ab Initio* Molecular Orbital Theory

A Born-Oppenheimer electronic wavefunction ϕ must obey the "clamped nuclei" Schrödinger equation described in Chapter 1,

$$h_e(\mathbf{r}|\mathbf{R})\phi(\mathbf{r}|\mathbf{R}) = E(\mathbf{R})\phi(\mathbf{r}|\mathbf{R}). \tag{A.1}$$

Here, h_e, the electronic Hamiltonian, might include spin-orbit operators as well as the usual kinetic energy, electron-nuclear, nuclear-nuclear, and electron-electron interaction terms. For any system containing more than one electron, equation A.1 has never been solved exactly, and one must resort to approximation methods to obtain a description of the wavefunction ϕ and a value for the electronic energy $E(\mathbf{R})$.

The two most commonly employed approximation techniques are perturbation theory (PT) and the variational method (VM) (Pilar, 1968; Eyring, Walter and Kimball, 1944). The implementation of either approximation begins with finding an appropriate set of molecular orbitals that can subsequently be used to construct a basis of N-electron functions in terms of which ϕ is expanded. Let us first analyze how the molecular orbitals are obtained.

A.1. Orbitals

In the Hartree-Fock (HF) or self-consistent field (SCF) method (Cook, 1978), one uses the variational principle to determine those spin-orbitals $\{\psi_i\}$—orbitals multiplied by a spin function α or β having $m_s = \pm 1/2$—that minimize the energy of a single Slater-determinant trial wavefunction ϕ_{SD} (Cook, 1978; Pilar, 1968)

$$\phi_{SD} = \det[\psi_1(\mathbf{r}_1)\psi_2(\mathbf{r}_2) \cdots \psi_N(\mathbf{r}_N)]. \tag{A.2}$$

This energy-minimization process results in a set of HF or SCF equations that the spin orbitals must obey, namely,

$$F\psi_1 = \epsilon_i \psi_i, \tag{A.3}$$

in which ϵ_i is the orbital energy corresponding to spin orbital ψ_i and F is the Fock operator in atomic units (Pilar, 1968), or

$$F = -\frac{1}{2}\nabla_r^2 - \sum_a \frac{Z_a}{|\mathbf{r} - \mathbf{R}_a|} + \sum_{j_{\text{occ}}}^N \int \psi_j^*(\mathbf{r}') \frac{1 - P_{rr'}}{|\mathbf{r} - \mathbf{r}'|} \psi_j(\mathbf{r}') \, d\mathbf{r}'. \tag{A.4}$$

in which $P_{rr'}$ permutes the coordinates $\mathbf{r}$ and $\mathbf{r}'$, and N is the number of electrons in the system. In the operator F, the sum over j_{occ} refers to those spin orbitals that are *occupied* in ϕ_{SD}. It is through this choice of occupancy that one determines the state (e.g., $1s\alpha 1s\beta$ or $1s\alpha 2s\beta$) for which the SCF calculation is being performed.

In writing the trial variational wavefunction as a single Slater determinant, one assumes that the major component of the true electronic wavefunction ϕ describes uncorrelated motion of the electrons. In other words, although the electrons certainly interact, their motion is not strongly affected by the *instantaneous* positions of the other electrons. The regions of space in which they move (the orbitals) are primarily determined by the *average* interactions among the electrons. This nearly independent-motion ansatz leads to the postulate that ϕ can be approximated as an antisymmetrized product of one-electron spin orbitals (Pilar, 1961; Cook, 1978); this approximation is similar to giving the probabilities of uncorrelated events as products of probabilities of the individual events.

If the orbitals used to construct ϕ_{SD} are allowed complete variational flexibility, the resulting calculation is referred to as an unrestricted HF (UHF) calculation. In this most general case, the resulting HF spatial orbitals associated with α and β spins will not necessarily be identical. For example, a UHF calculation on the $1s\alpha 1s'\beta 2s\alpha$ occupancy of Li does not yield two identical $1s$ orbitals ($1s \neq 1s'$). As a result, the Slater determinant $\det(1s\alpha 1s'\beta 2s\alpha)$ is not a pure doublet ($s = 1/2$) spin eigenfunction (Pauncz, 1979). Although this is indeed an unattractive feature of such UHF wavefunctions, this SCF procedure is widely used as a method for generating molecular orbitals (Pople, 1976). Subsequent to the UHF calculation of the molecular orbitals, the improper spin-symmetry behavior of ϕ_{SD} can be removed by applying a spin-projection operator P_s (Pauncz, 1979) to ϕ_{SD} to give a correct spin eigenstate

$$\phi_{s,\text{SD}} = P_s \phi_{\text{SD}}. \tag{A.5}$$

The resulting *projected* UHF wavefunction $\phi_{s,\text{SD}}$ is generally no longer a single Slater determinant. For example, a doublet ($s = 1/2$) projection of the $1s\alpha 1s'\beta 2s\alpha$ Slater determinant yields

$$\frac{1}{\sqrt{2}}[\det(1s\alpha 1s'\beta 2s\alpha) + \det(1s\beta 1s'\alpha 2s\alpha)].$$

However, it is still straightforward to compute the energy of $\phi_{s,\text{SD}}$ using the Slater-Condon rules (Cook, 1978) discussed below.

As an alternative to projecting a spin-unrestricted Slater determinant, one can force the orbitals that belong to paired electrons to be identical at the start of the SCF procedure and to remain so. For example, one can use the determinant $\det(1s\alpha 1s\beta 2s\alpha)$, which contains only two (1s and 2s) spatial orbitals rather than three (1s, 1s', and 2s). The energy of this spin-restricted HF (RHF) trial function can be minimized to give a set of equations (Roothaan, 1960) analogous to equation A.3, which determine the restricted HF orbitals. This RHF process has the advantage that it does not yield different spatial orbitals for paired electrons. Thus, for the Li example, the RHF ϕ_{SD} automatically has doublet spin symmetry (Pauncz, 1979). A disadvantage of the RHF method is that it is computationally more difficult. Furthermore, its derivation is not entirely free from arbitrary assumptions (Jørgensen and Simons, 1981), which makes it difficult to associate the orbital energies $\{\epsilon_j\}$ with ionization potentials by means of Koopman's theorem (Pilar, 1968; Cook, 1978). In contrast, the UHF method permits the approximate evaluation of (vertical) ionization energies as $-\epsilon_j$.

The UHF or RHF self-consistent-field equations are usually solved by the Roothaan-matrix procedure in which the ψ_i functions are expanded in an atomic orbital (AO) basis $\{X_b\}$. When this expansion is used in equation A.3, one obtains Roothaan-matrix HF equations of either the UHF or RHF variety (Cook, 1978). If M atomic orbitals are used in the expansion, the resulting matrix eigenvalue problem generates N *occupied* molecular orbitals and $2M - N$ excited or *virtual* molecular orbitals.

$$\psi_i = \sum_b C_{ib} X_b (\alpha \text{ or } \beta). \tag{A.6}$$

The most commonly used atomic orbitals are Slater-type orbitals (STO), namely,

$$X_b = N_b Y_{l_b m_b} r^{n_b - 1} \exp(-\zeta_b |\mathbf{r} - \mathbf{R}_b|) \tag{A.7}$$

and Gaussian-type orbitals (GTO),

$$X_a = N'_a X^{k_a} Y^{u_a} Z^{v_a} \exp[-\alpha_a (\mathbf{r} - \mathbf{R}_a)^2]. \tag{A.8}$$

In these defining equations, Y_{lm} is a spherical harmonic, N_b and $N_{a'}$ are normalization constants, $R_{a,b}$ is the position of the nucleus on which the atomic orbital is located, n_b, l_b, m_b, k_a, u_a and v_a are orbital quantum numbers, and ζ_b and α_a are orbital exponents that determine the radial sizes of the atomic orbital.

Slater-type orbitals are to be preferred on fundamental grounds because they display proper cusp behavior at the nuclear centers. For example, the slope of a 1s Slater-type orbital at the nucleus is related to the orbital exponents

$$\left[\frac{d}{dr} N \exp(-\zeta r)\right]_{r=0} = -N\zeta. \tag{A.9}$$

This is precisely the behavior displayed by hydrogenlike orbitals that are eigenfunctions of the one-electron Schrödinger equation having only kinetic and electron-nuclear attraction energies. In contrast, all GTO have zero slope at the nucleus; for example, for the 1s GTO

$$\left[\frac{d}{dr} N \exp(-\alpha r^2)\right]_{r=0} = 0. \tag{A.10}$$

Near a nucleus the full Schrödinger differential equation is dominated by the same kinetic and nuclear-electron attraction terms that constitute the hydrogenlike Hamiltonian; thus, the correct wavefunction ϕ must display hydrogenlike cusps at the nuclei. The STO fulfill this criterion; the GTO do not.

The deficiencies of the GTO raise the question of why and how they are used. GTO's are convenient in studies of polyatomic molecules because they allow efficient handling of the multicenter integrals that arise. In integrals containing a product of two orbitals X_a and X_b that have origins on different nuclei $\mathbf{R}_a$ and $\mathbf{R}_b$, the Gaussian orbitals allow this product to be written in terms of a single common origin. For example, the product of two 1s-type GTO's can be expressed as

$$\exp[-\alpha_a(\mathbf{r}-\mathbf{R}_a)^2]\exp[-\alpha_b(\mathbf{r}-\mathbf{R}_b)^2]$$
$$= \exp[-(\alpha_a+\alpha_b)\mathbf{r}^2]\exp\left(-\frac{\alpha_a\alpha_b}{\alpha_a+\alpha_b}\mathbf{R}^2\right) \tag{A.11}$$

in which the origin of the final r-dependent function is located between $\mathbf{R}_a$ and $\mathbf{R}_b$ at a distance $\alpha_a R/(\alpha_a+\alpha_b)$ from $\mathbf{R}_a$ and $\alpha_b R/(\alpha_a+\alpha_b)$ from $\mathbf{R}_b$ ($R \equiv |\mathbf{R}_a - \mathbf{R}_b|$). The fact that the product $X_a X_b$ that involves GTO's having different origins can be expressed as a single new GTO at a new origin makes the use of GTO's in evaluating integrals efficient.

To attempt to overcome the improper cusp behavior of GTO's, one often employs *contracted* GTO's (CGTO) (Schaefer, 1972; Dunning, 1970, 1971; Huzinaga, 1965). A CGTO (X_a^c) is a linear combination of the GTO's

$$X_a^c = \sum_b A_{ab} X_b \tag{A.12}$$

in which the GTO $\{X_b\}$ have common quantum numbers (e.g., 1s, 2p, 3d) but different orbital exponents (α_b). By combining a tight GTO (one having large

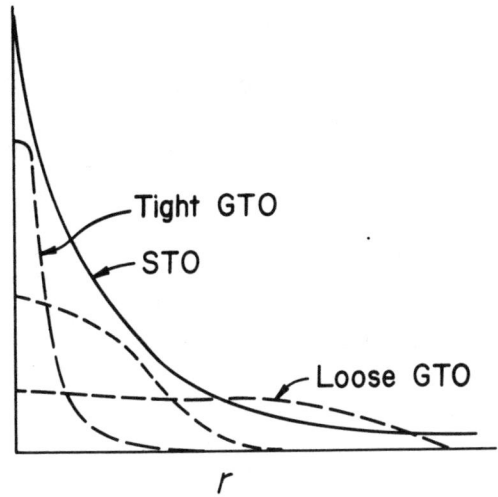

Figure A-1
Formation of an STO by addition of loose and tight GTO's.

α_b) with other GTO's having progressively smaller exponents, one might fit the cusp behavior of the STO. This is shown for $1s$-type functions in Figure A-1. By choosing the contraction coefficients $\{A_{ab}\}$ properly, it is possible to generate a CGTO which, in a least-squares sense, reproduces the proper STO cusp behavior (Pople, 1969). Alternatively, the contraction coefficients can be chosen to minimize the SCF energy of the lowest energy state of the atom of interest (Dunning, 1970, 1971). For either of these CGTO for any atom, the CGTO itself can be viewed as the atomic-orbital basis function that is to be used in subsequent molecular SCF calculations. In the literature, tabulations of optimal CGTO's are available for most first-, second-, and third-row atoms based either upon the STO fitting procedure (Pople, 1969) or the atomic-energy optimization procedure (Dunning, 1970, 1971).

In summary, the SCF method can be used to generate a set of molecular orbitals that are expressed in terms of a chosen set of Gaussian- or Slater-type atomic basis functions. The nature of the occupied molecular orbitals, as displayed in the molecular-orbital expansion coefficients of equation A.6, describes the charge density and bonding characteristics of these orbitals. The energies $\{\epsilon_i\}$ of the occupied orbitals give us, via Koopmans' theorem, ionization potentials of the system. However, one must remember that the entire SCF method, including the concept of molecular orbital, is predicated upon the *assumption* that ϕ is accurately represented by ϕ_{SD}.

A.2. Configuration Interaction

In Chapter 3 it was shown that it is not always possible to describe the electronic wavefunction ϕ in terms of a *single* orbital-occupancy list (configuration). For example, the fragmentation of heteropolar bonds to give radical products was shown to require both σ^2 and $\sigma\sigma^*$ configurations to describe ϕ throughout the entire bond-dissociation process. For these reasons one must often extend the description of ϕ to include more than one Slater determinant or configuration. Moreover, such a multiconfigurational description should be examined whenever high accuracy in the resulting wavefunction and energy is desired. Hence, even when ϕ is dominated ($\sim 95\%$) by a single Slater determinant, that determinant does not accurately represent the true wavefunction because the Slater determinant describes electrons moving in orbitals determined only by the average interactions with the other electrons and not by the instantaneous interactions. No electron correlation is present in the SCF (single determinant) description.

By writing ϕ as a linear combination of all Slater determinants $\{\phi_I\}$ that can be constructed from the $2M$ (occupied and virtual) SCF spin orbitals,

$$\phi = \sum_I C_I \phi_I, \tag{A.13}$$

the SCF treatment is improved. In most such configuration-interaction calculations the ϕ_I are symmetry-projected functions (each of which may contain several Slater determinants) describing the various configurations (spin-orbital occupancies) that can be made from $2M$ orbitals and N electrons (Pauncz, 1979; Shavitt, 1977).

Clearly, the number $2M!/N!(2M - N)!$ of these configurations becomes extremely large as the basis size (M) and the number of electrons increases. Therefore, various procedures have evolved for selecting the most important of the $2M!/N!(2M - N)!$ configurations (Shavitt, 1977). The most commonly used criterion for judging the importance of a configuration ϕ_I is to evaluate its interaction strength using the one or few configurations that are absolutely essential to describe ϕ. For example, for the case of heteropolar bond rupture (mentioned above and in Chapter 3) the σ^2 and $\sigma\sigma^*$ configurations are essential. Other configurations ϕ_I (e.g., $\sigma\sigma'$, π^2, and so forth) are evaluated for their importance by computing their interaction strengths $\langle \phi_I | H | \phi_{\text{essential}} \rangle$ for all of the essential configurations. If the interaction strength is large, ϕ_I is considered to be important.

Unfortunately, ranking configurations according to the value of their interaction strengths provides little physical interpretation for the nonessential configurations. However, it is possible to ascribe meaning to those configurations that are either singly (ϕ_s) or doubly (ϕ_D) excited relative to a dominant essential configuration ϕ_e. The contributions to ϕ made by a configuration in which one electron is promoted from ψ_μ (occupied in ϕ_e) to ψ_p (unoccupied in ϕ_e) can be denoted $\phi_{s\mu}^p$. The trial wavefunction $\phi_e + C_s\phi_{s\mu}^p$ consisting of two Slater determinants is (because the two determinants differ by one column only) equivalent to another *single* Slater determinant having the property that the column in which ϕ_e and $\phi_{s\mu}^p$ differ (ϕ_e having ψ_μ and $\phi_{s\mu}^p$ having ψ_p) is replaced by a column containing the *modified* spin orbital $\psi' = \psi_\mu + C_s\psi_p$. The fact that $\phi_e + C_s\phi_{s\mu}^p$ is equivalent to a modified Slater determinant in which ψ_μ has been replaced by ψ' is the basis for saying that such singly excited determinants $\psi_{s\mu}^p$ produce either orbital modification or orbital relaxation.

If the SCF orbitals are used to construct the Slater determinants, one finds (by using the Slater-Condon rules discussed below) that the interaction strength between the SCF determinant (which is presumably one of the essential configurations) and singly excited determinants vanishes—that is,

$$\langle \phi_{\text{HF}} | H | \phi_s \rangle = 0. \tag{A.14}$$

This equation, known as the Brillouin theorem (Schaefer, 1972; Cook, 1978; Pilar, 1968), simply states that singly excited determinants are not important (in the interaction strength case) when SCF orbitals are used to construct the determinants because these orbitals are already optimal—further optimization (modification or relaxation) or the orbitals is not needed.

For doubly excited configurations $\phi_{D,\mu\nu}^{pq}$ in which ψ_μ and ψ_ν (which are occupied in ϕ_e) are replaced by ψ_p and ψ_q, the trial function $\phi_e - C_D\phi_{D,\mu\nu}^{pq}$ can be rewritten as a combination of two other determinants, each of which include *pairs* of polarized orbitals:

$$\phi_e - C_D\phi_{D,\mu\nu}^{pq} = \det[\ldots(\psi_\mu - \sqrt{C_D}\,\psi_p)(\psi_\nu + \sqrt{C_D}\,\psi_q)]$$
$$+ \det[\ldots(\psi_\mu + \sqrt{C_D}\,\psi_p)(\psi_\nu - \sqrt{C_D}\,\psi_q)]. \tag{A.15}$$

Notice that in each of these two determinants, each electron moves in separate polarized orbitals. For example, in describing the rupture of the H_2 bond discussed in Chapter 3, a configuration-interaction wavefunction including both the σ_g^2 and σ_u^2 configurations was used:

$$\phi = \sigma_g^2 - C\sigma_u^2. \tag{A.16}$$

The polarized orbitals corresponding to this doubly excited configuration-interaction wavefunction are $\sigma_g \pm \sqrt{C}\sigma_u$. Such double excitations give rise to electron-pair correlations because, in the polarized-orbital determinantal description, the electron pair residing in the orbitals ψ_μ and ψ_ν are correlated in the sense that one electron is in one polarized orbital when the other electron is

Figure A-2
Polarized orbitals.

in the other polarized orbital. Doubly excited determinants that include excitations of orbitals differing in their left-right symmetry (as above) give rise to polarized orbitals that are left-right polarized. Double excitations of the form $\sigma^2 \to \pi_x^2$ give rise to polarized orbitals ($\sigma \pm \sqrt{C}\pi_x$) that differ in their angular characteristics as shown in Figure A-2. Double excitations of the form $2s^2 \to 3s^2$ give polarized orbitals ($2s \pm \sqrt{C}3s$) that differ in their radial or in-out character. This polarized orbital-*pair* description of the contributions made by double excitations is the basis for saying that such configurations give rise to electron correlations. As one electron is in one polarized orbital, the second is in the other orbital; this is what is meant by correlated motion.

In summary, configuration interaction is used to improve upon the SCF description of the electronic wavefunction. Such improvement is often essential as, for example, in describing heteropolar bond rupture to give radical products. However, configuration interaction can also be used simply to improve the accuracy of the wavefunction ϕ and energy E. Configurations that are singly or doubly excited relative to a dominant (essential) configuration allow orbital relaxation and electron-pair correlation effects to be included in the configuration-interaction wavefunction. Numerical procedures for adding configurations above and beyond these singles and doubles, which are included on physical grounds, are usually based upon evaluating the interaction strength of each such configuration with all of the essential configurations.

A.3. Slater-Condon Rules

After obtaining a list of configurations that includes the essential configurations and perhaps some set of singly, doubly, or more highly excited configurations that have been chosen as discussed in section A.2, the C_I expansion coefficients of equation A.13 must be evaluated. In the configuration-interaction procedure, the wavefunction of equation A.13 is used in the variational method to minimize the electronic energy. This approach leads to the well-known configuration-interaction matrix eigenvalue problem (Shavitt, 1977; Pilar, 1968)

$$\sum_{J=1}^{Q} H_{IJ} C_J = E C_I \qquad I = 1, \ldots Q \tag{A.17}$$

in which Q is the number of configurations included in the configuration-interaction wavefunction and E is the configuration-interaction approximation to the electronic energy. The $Q \times Q$ matrix **H** clearly has Q eigenvalues $\{E_i, i = 1, \ldots Q\}$ and Q independent eigenvectors $\{C_{i,J}, i = 1, \ldots Q; J = 1 \ldots Q\}$. The various E_i represent the configuration-interaction approximations to the ground- and excited-state energies; the coefficients $\{C_{i,J}; J = 1, \ldots Q\}$ describe the configuration-interaction wavefunction for this ith state.

The elements of the **H** matrix are given as integrals over the configurations ϕ_I:

$$H_{IJ} = \int \phi_I^* h_e \phi_J \, d\tau_1 \ldots d\tau_N, \tag{A.18}$$

in which h_e is the full Born-Oppenheimer electronic Hamiltonian described earlier and $d\tau_j$ denotes the space- and spin-integration volume element for the jth electron. The evaluation of these integrals is nontrivial because the ϕ_I are antisymmetrized N-electron functions. The derivation of closed expressions for the $H_{I,J}$ matrix elements is given in many texts on quantum chemistry (Cook, 1978; Condon and Shortley, 1957). The resulting set of so-called Slater-Condon rules can be summarized as follows. Two configuration functions ϕ_I and ϕ_J are first decomposed into their constituent Slater determinants $\det_{I_k}$ and $\det_{J_l}$ (each ϕ_I consists of one or more Slater determinants). To compute the matrix element $\langle \det_{I_k} | h_e | \det_{J_l} \rangle$ the spin-orbital occupancies of these two determinants are compared. If the occupancies differ by more than two spin orbitals (e.g., $1s^2 2s^2$ and $1s_\alpha 2p_x^2 3s_\beta$ differ by three), then the matrix element vanishes. If the occupancies differ by two spin orbitals (with ψ_μ and ψ_ν in $\det_{I_k}$ and ψ_p and ψ_q in $\det_{J_l}$), then the matrix element has a value $\pm [\langle \mu\nu | pq \rangle - \langle \mu\nu | qp \rangle]$ in which

$$\langle \mu\nu | pq \rangle \equiv \int \psi_\mu^*(\mathbf{r}) \psi_\nu^*(\mathbf{r}') |\mathbf{r} - \mathbf{r}'|^{-1} \psi_p(\mathbf{r}) \psi_q(\mathbf{r}') \, d\tau d\tau'. \tag{A.19}$$

The choice of $+$ or $-$ in the $\pm$ sign is determined by how many spin-orbital interchanges are needed to arrange $\det_{J_l}$ to have exactly the same spin-orbital *ordering* as $\det_{I_k}$, except that ψ_p replaces ψ_μ and ψ_q replaces ψ_ν. If the number of interchanges needed is odd (even), then the minus (plus) sign results. When $\det_{I_k}$ and $\det_{J_l}$ differ by only one spin orbital (with ψ_μ in $\det_{I_k}$ and ψ_p in $\det_{J_l}$), then the value of the matrix element is

$$\pm \left[\langle \psi_\mu | -\frac{1}{2}\nabla_r^2 - \sum_a \frac{Z_a}{|\mathbf{r}-\mathbf{R}_a|} | \psi_p \rangle + \sum_\nu (\langle \psi_\mu \psi_\nu | \psi_p \psi_\nu \rangle - \langle \psi_\mu \psi_\nu | \psi_\nu \psi_p \rangle) \right]$$

in which the sum over ν runs over all of the spin orbitals common to $\det_{I_k}$ and $\det_{J_l}$. The sign $\pm$ is computed as just described by determining how many spin-orbital interchanges are needed to bring $\det_{J_l}$ into the same order (except for the ψ_p and ψ_μ mismatch) as $\det_{I_k}$. If $\det_{I_k}$ and $\det_{J_l}$ have identical spin-orbital occupancies, their Hamiltonian matrix element is given as above, but with $\psi_p = \psi_\mu$, and summed over the index μ that runs over all occupied spin orbitals.

Once the Slater-Condon rules are used to compute the Hamiltonian matrix elements over the determinants $\det_{I_k}$, the evaluation of the configuration-based matrix elements $\langle \phi_I | h_e | \phi_J \rangle$ is straightforward. Knowing that ϕ_I is expressed as a linear combination of the $\det_{I_k}$

$$\phi_I = \sum_k B_{I_k} \det_{I_k}, \tag{A.20}$$

one can write

$$\langle \phi_I | h_e | \phi_J \rangle = \sum_{k,l} B_{I_k} B_{J_l} \langle \det_{I_k} | h_e | \det_{J_l} \rangle. \tag{A.21}$$

The final result of using the Slater-Condon rules is that the configuration-interaction **H** matrix, whose dimension is equal to the number of configurations selected, can be computed in terms of the one-electron ($\langle \psi_i | -1/2\nabla^2 | \psi_j \rangle$, $\langle \psi_i | -Z_a/|\mathbf{r}-\mathbf{R}_a| | \psi_j \rangle$) and two-electron ($\langle \psi_i \psi_j | \psi_k \psi_l \rangle$) integrals over the spin orbitals used to form the configurations. These integrals can be evaluated in terms of the molecular-orbital expansion coefficients C_{ia} and the one- and two-electron integrals over the atomic-basis orbitals $\{X_a\}$, which must be explicitly calculated for the GTO or STO basis. For example, the two-electron integrals are expressed as

$$\langle \psi_i \psi_j | \psi_k \psi_l \rangle = \sum_{abcd} C_{ia} C_{jb} C_{kc} C_{ld} \langle X_a X_b | X_c X_d \rangle. \tag{A.22}$$

After forming the $Q \times Q$ configuration-interaction **H** matrix, the eigenvalues (E_I) and eigenvectors (C_{IJ}; $J = 1 \ldots Q$) are found by diagonalization. Each of the resulting approximate energy levels E_I can be shown (Hylleraas, 1930) to be an upper bound to the Ith exact energy level of the system. The eigenvector $\{C_{IJ}; J = 1 \ldots Q\}$ tell us to express the approximate configuration-interaction wavefunction (ϕ_I) for the Ith energy level in terms of the configurations $\{\phi_J\}$

$$\phi_i = \sum_{I=1}^{Q} C_{iI}\phi_I. \tag{A.23}$$

By including in a configuration-interaction wavefunction the essential configurations (which are usually straightforward to guess) as well as configurations that are singly and doubly excited relative to any of these essential components, a satisfactory description of orbital relaxation and electron-pair correlation effects can often be achieved. This truncated configuration-interaction treatment, based upon low-order excitations out of essential configurations, has a significant weakness that should be made clear. This kind of configuration-interaction method suffers from what is called *size inconsistency* (Pople, 1976). To illustrate the problem, consider how one would compute the configuration-interaction energy of two separated and noninteracting beryllium atoms. Assume that a configuration-interaction calculation has already been performed on a single Be atom from which it was decided that only two configurations ($1s^2 2s^2$ and $1s^2 2p^2$) need to be included in the one-atom configuration-interaction wavefunction to achieve a reasonable description—that is, evidence is available that supports the inclusion of only double excitations ($2s^2 \rightarrow 2p^2$) in the Be atom configuration-interaction wavefunctions. If the *same* level of configuration interaction (dominant configuration $1s_A^2 2s_A^2 1s_B^2 2s_B^2$ plus double excitations $2s_A^2 \rightarrow 2p_A^2$, $2s_A^2 \rightarrow 2p_A 2p_B$, $2s_B^2 \rightarrow 2p_B^2$, $2s_A 2s_B \rightarrow 2p_A 2p_B$, $2s_A 2s_B \rightarrow 2p_A$, and $2s_A 2s_B \rightarrow 2p_B^2$) were applied to the Be + Be system, the lowest resultant configuration-interaction energy would *not* be equal to twice the configuration-interaction energy obtained above for the single Be atom. One says that this configuration-interaction treatment is *size-inconsistent* because the energy obtained for noninteracting systems is not the sum of the configuration-interaction energies of the individual systems.

What is wrong with the above configuration-interaction wavefunction is that the wavefunction for Be + Be *should* (because the atoms are noninteracting) be the antisymmetrized product of the wavefunctions for the two Be atoms (A and B):

$$\phi_{Be + Be} = \phi_{Be_A}\phi_{Be_B}. \tag{A.24}$$

Because ϕ_{Be_A} and ϕ_{Be_B} contain both $1s^2 2s^2$ and $1s^2 2p^2$ configurations, $\phi_{Be + Be}$ should contain $2s_A^2 2s_B^2$, $2s_A^2 2p_B^2$, $2p_A^2 2s_B^2$, *and* $2p_A^2 2p_B^2$ (the $1s_A^2 1s_B^2$ is suppressed). This last configuration is quadruply excited relative to the dominant $2s_A^2 2s_B^2$ configuration, but it must be included if $\phi_{Be + Be}$ is to be size-consistent.

From the above example, it should be clear that a configuration-interaction wavefunction that is truncated to any level of excitation (e.g., doubly) when separately treating individual systems, A and B, will not be appropriate for use when treating the combined system AB even when A and B are far removed (let alone when they are interacting or chemically bonded).

Correct treatment of AB requires inclusion of excitations up through the sum of the excitation levels used when separately treating A and B. Clearly, this size-consistency problem of the truncated configuration-interaction method may cause serious errors when using these techniques for computing energy differences such as bond-dissociation energies, intermolecular forces, and energy or enthalpy changes in chemical reactions.

The realization that truncated configuration-interaction approximations are not size-consistent has led to much recent interest in the use of perturbation theory for treating electron correlation effects (Pople, 1976). In these many-body perturbation theories (MBPT) the electronic Hamiltonian h_e is usually decomposed into h_e^0 and V in which h_e^0 is a sum of one-electron Fock operators

$$h_e^0 = \sum_{i=1}^{N} F(i), \tag{A.25}$$

in which $F(i)$ is the Fock operator for the ith electron defined in equation A.4. The perturbation V then consists of the instantaneous electron-electron interaction minus the average (coulomb minus exchange) interaction contained in the Fock operator

$$V = \frac{1}{2}\sum_{i \neq j}^{N} \frac{1}{|\mathbf{r}_i - \mathbf{r}_j|} - \sum_{i=1}^{N} v(\mathbf{r}_i) \tag{A.26}$$

in which

$$v(\mathbf{r}_i) = \sum_{j_{occ}} \int \psi_j^*(\mathbf{r}') \frac{1 - P_{r_i, r'}}{|\mathbf{r}_i - \mathbf{r}'|} \psi_j(\mathbf{r}') \, d\mathbf{r}'. \tag{A.27}$$

In addition to the above decomposition of h_e, the exact electronic wavefunction ϕ is assumed to be given as a zeroth-order component ϕ^0 plus higher-order corrections ($\phi^{(n)}$; $n = 1, 2, \ldots$) with ϕ^0 taken to be a *single-*configuration wavefunction. Such a single-configuration wavefunction, if it is constructed from the SCF spin orbitals, is an eigenfunction of the above-defined h_e^0 with eigenvalue E^0 equal to the sum of the SCF orbital energies belonging to the spin orbitals occupied in ϕ^0.

Although such MBPT-based treatments of electron correlations have been successfully carried out by several research workers (Pople, 1976; Bartlett, 1975), major problems arise when the physical situation dictates that the true wavefunction ϕ is not dominated by a *single* configuration. As we saw earlier, description of processes with heteropolar or homopolar bond rupture usually requires a description of two or more essential configurations. By using a spin-unrestricted SCF configuration as ϕ^0, it is possible to describe bond rupture within the single-configuration picture. Near the equilibrium bond length of, length of, for example, HCl, the single configuration is det$(\ldots \sigma\alpha\sigma'\beta)$ with $\sigma = \sigma'$. Upon bond rupture the single-configuration wavefunction becomes det$(\ldots \sigma\alpha\sigma'\beta)$ in which σ is the hydrogen $1s$ orbital and σ' is a chlorine $3p$ orbital. Unfortunately, such a UHF treatment suffers from the spin-impurity difficulty discussed in section A.1. Moreover, chemical problems (e.g., strong configuration mixing arising in concerted reactions involving breaking and forming more than one bond) exist for which *any* single-configuration description is inappropriate, and these limit the application of MBPT to large numbers of species arising in a variety of chemical reactions.

At present, a great deal of research is aimed at extending the machinery of MBPT (which does not suffer from the size-consistency problems) to permit ϕ^0 to consist of more than one essential configuration; however, this problem is not resolved yet. As a result, a perturbation theory tool that can be used in the avoided-crossing situations arising in many of the concerted reactions treated in Chapters 4 and 7 is not available. Thus, we shall not pursue further the use of MBPT to treat correlation in a manner that overcomes the size-consistency difficulty of the configuration-interaction method. The most essential point is that configuration interaction is not size-consistent. Thus, although the inclusion of singly and doubly excited configurations is attractive because of their significance with respect to relaxation and pair correlation, use of the variational configuration-interaction method for determining the *amplitudes* of these configurations may be questionable. Unfortunately, the MBPT method has not yet been extended to allow multiconfigurational zeroth-order functions, so it also cannot (at present) be employed for reliable evaluation of the desired amplitudes.

Appendix B

The Nature of Photon-Induced Electronic Transitions

When a photon is absorbed by a molecule and causes an electronic transition to occur, the *electronic* energy of the molecule changes from $E_0(\mathbf{R})$, its ground-state value before absorption, to $E_x(\mathbf{R})$, its excited-state value. The energy of the photon of frequency $h\nu$ must match $E_x(\mathbf{R}) - E_0(\mathbf{R}) = h\nu$. For any given frequency ν, this condition will generally not be met at *all* molecular geometries $\{\mathbf{R}\}$; only at particular geometries $\{\mathbf{R}_c\}$ will $h\nu = E_x - E_0$.

There is much more to understanding photon absorption than is contained in the above relation. Often the electronic absorption spectrum of a molecule (even one which subsequently undergoes a photoreaction) displays sharp vibrational structure, especially when the molecule is in the gas phase or in an inert matrix such as frozen argon or nitrogen. This vibrational structure arises because the ground and excited electronic states of the molecule have quantized vibrational energy levels $\{\epsilon_v^0\}$ and $\{\epsilon_{v'}^x\}$, respectively. Even when the excited state has vibrational levels that are broadened by dissociation (i.e., they are not actually bound), vibrational structure can persist in the absorption spectrum if the width of the state ($\hbar$ divided by the dissociation lifetime) remains less than the spacing between the levels.

When vibrational structure is seen, the energy of the photon must also obey the equation $h\nu = \epsilon_{v'}^x - \epsilon_v^0$; that is, the transition occurs between quantized states of E_0 and E_x. Combining the above two requirements on $h\nu$ gives

$$\epsilon_{v'}^x - \epsilon_v^0 = E_x(\mathbf{R}_c) - E_0(\mathbf{R}_c). \tag{B.1}$$

This very important relation shows that transitions from ϵ_v^0 to $\epsilon_{v'}^x$ can occur only at molecular geometries $\{\mathbf{R}_c\}$ in which the quantum-level energy difference $\epsilon_{v'}^x - \epsilon_v^0$ is identical to the *electronic* energy difference $E_x(\mathbf{R}_c) - E_0(\mathbf{R}_c)$. Since the electronic energy functions E_x and E_0 are the potential energy functions for the vibration-rotation motion of the molecule, the vibration-rotation classical kinetic energies T are given by $\epsilon_{v'}^x - E_x(\mathbf{R}_c) = T_x(\mathbf{R}_c)$ and $\epsilon_v^0 - E_0(\mathbf{R}_c) = T_0(\mathbf{R}_c)$. Hence, the above condition can be restated as $T_x(\mathbf{R}_c) = T_0(\mathbf{R}_c)$; that is, photon absorption can occur at geometries in which the classical vibration-rotation kinetic energy is conserved.

Thus far, we have seen how to determine geometries at which the light of energy $h\nu = \epsilon_{v'}^x - \epsilon_v^0$ can be absorbed. To understand the rate at which such light *will* be absorbed, we need to consider what happens to the electronic and vibration-rotation wavefunctions of the molecule when a photon is absorbed.

In the approximation in which the photon-molecule interaction is treated as an electric dipole interaction, the ground-state Born-Oppenheimer wavefunction $\phi_0(\mathbf{r}|\mathbf{R})\chi_v^0(\mathbf{R})$ becomes $\phi_x(\mathbf{r}|\mathbf{R})\langle\phi_x(\mathbf{r}|\mathbf{R})|\boldsymbol{\epsilon}\cdot\mathbf{r}|\phi_0(\mathbf{r}|\mathbf{R})\rangle\chi_v^0(\mathbf{R})$ when the photon ($h\nu = \epsilon_{v'}^x - \epsilon_v^0$) is absorbed (Simons, 1982). To make a connection with the Franck-Condon picture, the **R**-dependence of the above postabsorption wavefunction ($\psi_{x,\nu}$) is expanded in terms of the complete set of vibration-rotation functions $\{\chi_{v'}^x\}$ of the excited state

$$\psi_{x,\nu}(\mathbf{r}|\mathbf{R}) = \sum_{v'}\langle\chi_{v'}^x\phi_x|\boldsymbol{\epsilon}\cdot\mathbf{r}|\chi_v^0\phi_0\rangle\phi_x\chi_{v'}^x\delta\left[\nu - \left(\frac{\epsilon_{v'}^x - \epsilon_v^0}{h}\right)\right] \quad (B.2)$$

The δ function is inserted simply to insure that the energy of the photon that creates $\psi_{x,\nu}$ is equal to the quantum-state energy difference $\epsilon_{v'}^x - \epsilon_v^0$. The probability P of finding the molecule in any specific vibration-rotation state $\phi_x\chi_{v'}^x$ having energy $\epsilon_{v'}^x$ is given by the square of the amplitude of this state in the above expression for $\psi_{x,\nu}$:

$$P = \delta\left[\nu - \left(\frac{\epsilon_{v'}^x - \epsilon_v^0}{h}\right)\right]|\langle\chi_{v'}^x\phi_x|\boldsymbol{\epsilon}\cdot\mathbf{r}|\chi_v^0\phi_0\rangle|^2. \quad (B.3)$$

The usual Franck-Condon factors arise by assuming that the electric-dipole-transition matrix element $\langle\phi_x(\mathbf{r}|\mathbf{R})|\boldsymbol{\epsilon}\cdot\mathbf{r}|\phi_0(\mathbf{r}|\mathbf{R})\rangle \equiv \mu_{0x}(\mathbf{R})$ is relatively independent of molecular geometry $\mu_{0x}(\mathbf{R}) \cong \mu_{0x}$. With such an approximation

$$P = \delta\left[\nu - \frac{\epsilon_{v'}^x - \epsilon_v^0}{h}\right]|\langle\chi_{v'}^x|\chi_v^0\rangle|^2\mu_{0x}^2,$$

which contains the Franck-Condon factors $|\langle\chi_{v'}^x|\chi_v^0\rangle|^2$. The expression for P leads to the conclusion that transitions to $\phi_x\chi_{v'}^x$ occur at a rate proportional to μ_{0x}^2 times the square of the overlap between the initial vibration-rotation state χ_v^0 and the final state $\chi_{v'}^x$.

Transitions for which μ_{0x} vanish are said to be electronically forbidden. Molecular point-group symmetry, which is reflected in the spatial symmetry of ϕ_0 and ϕ_x, can determine whether μ_{0x} vanishes. For example, the $^1A_1 \rightarrow {}^1A_2$, $n(b_2) \rightarrow \pi^*(b_1)$ electronic transition in C_{2v} H$_2$CO is forbidden, since $\mu_{0x} = \langle\pi^*|\boldsymbol{\epsilon}\cdot\mathbf{r}|n\rangle$ vanishes. Transitions that are electronically allowed can still be forbidden if the Franck-Condon factor $|\langle\chi_{v'}^x|\chi_v^0\rangle|^2$ vanishes. Again, molecular symmetry gives rise to symmetry in $\chi_{v'}^x$ and χ_v^0, which can then be used to predict whether $\langle\chi_{v'}^x|\chi_v^0\rangle$ vanishes. In special circumstances it is possible for electronically forbidden transitions to give rise to weak absorption intensities. If in equation B.3 it had not been assumed that $\mu_{0x}(\mathbf{R})$ is $\mathbf{R}$-independent, and $\mu_{0x}(\mathbf{R})$ was expanded about the equilibrium geometry $\{\mathbf{R}_e\}$ of the ground state χ_ν^0,

$$\mu_{0x}(\mathbf{R}) = \mu_{0x}(\mathbf{R}_e) + (\mathbf{R} - \mathbf{R}_e) \cdot \nabla_R \mu_{0x} + \cdots, \tag{B.4}$$

then for such electronically forbidden transitions the transition probability P would reduce to

$$P = \delta\left[\nu - \left(\frac{\epsilon_{\nu'}^x - \epsilon_\nu^0}{h}\right)\right] |\nabla_R \mu_{0x} \cdot \langle \chi_{\nu'}^x | (\mathbf{R} - \mathbf{R}_e) | \chi_\nu^0 \rangle|^2. \tag{B.5}$$

Because the geometrical displacements $\mathbf{R} - \mathbf{R}_e$ contain contributions from various symmetries, the integrals $\langle \chi_{\nu'}^x | \mathbf{R} - \mathbf{R}_e | \chi_\nu^0 \rangle$ could be nonzero even though $\langle \chi_{\nu'}^x | \chi_\nu^0 \rangle = 0$. In such cases, the intensity of the transition is said to be *borrowed*. In lowest order it is forbidden, since $\mu_{0x} \cong 0$; it is only through the $\mathbf{R}$-dependence of μ_{0x} that the transition is weakly allowed.

Although the above Franck-Condon analysis of the intensities of vibration-rotation structure in electronic absorption lines is very informative, another point of view gives additional insight. By treating the vibration-rotation kinetic-energy operator of the molecule classically, the photon-absorption probability can be rewritten as follows (Simons, 1982):

$$P = \langle \chi_\nu^0 | \delta[\nu - (E_x(\mathbf{R}) - E_0(\mathbf{R}))/h] \mu_{0x}^2(\mathbf{R}) | \chi_\nu^0 \rangle. \tag{B.6}$$

This expression can be interpreted in terms of the probability $|\chi_\nu^0(\mathbf{R})|^2$ of the molecule being at geometry $\mathbf{R}$ in the ground state χ_ν^0, multiplied by the electric dipole matrix element at that geometry $\mu_{0x}^2(\mathbf{R})$, and constrained (by the δ function) to allow contribution of only those geometries that obey $h\nu = E_x - E_0$. By allowing equation B.6 to apply only when $h\nu = \epsilon_{\nu'}^x - \epsilon_\nu^0$, a partly classical approximation of P is obtained:

$$P = \delta[\nu - (\epsilon_{\nu'}^x - \epsilon_\nu^0)/h] \langle \chi_\nu^0 | \delta[\nu - (E_x - E_0)/h] \mu_{0x}^2 | \chi_\nu^0 \rangle. \tag{B.7}$$

This expression for P can be used in attempting to understand how photon absorption prepares the molecule at the excited-state potential-energy surface $E_x(\mathbf{R})$. The energy of the photon must coincide (within the spectral linewidths) with one of the energy spacings $\epsilon_{\nu'}^x - \epsilon_\nu^0$. For each such energy value, the molecule can absorb the light only at geometries $\{\mathbf{R}_c\}$ obeying $h\nu = E_x(\mathbf{R}_c) - E_0(\mathbf{R}_c)$; this condition preserves the vibration-rotation kinetic energy of the molecule. The relative probability that the molecule experiences each such critical geometry $\{\mathbf{R}_c\}$ is given by the square of the initial vibration-rotation wavefunction $|\chi_\nu^0(\mathbf{R}_c)|^2$.

The overall relative probability of finding the molecule on the excited surface E_x at any $\mathbf{R}_c$ for a given photon energy $h\nu = \epsilon_{\nu'}^x - \epsilon_\nu^0$ is the probability

$|\chi_v^0|^2$ of the absorbing molecule being at $\{\mathbf{R}_c\}$ multiplied by the relative rate $\mu_{0x}^2(\mathbf{R}_c)$ of its electronic absorption at $\{\mathbf{R}_c\}$. This interpretation of equation B.7 is a valuable one. In attempting to determine the geometries at which the system will enter the excited surface, only molecular geometries for which $|\chi_v^0|^2$ is substantial must be examined. Within such geometries, only those for which ground- and excited-state surfaces are spaced by $h\nu$ will be populated during photon absorption. Finally, the transition to the excited surface will be efficient only where $\mu_{0x}^2(\mathbf{R}_c)$ is large.

The above qualitative treatment of photon absorption was motivated by the need to guess where a molecule will enter an excited-state potential energy surface. Knowing where it enters E_x, one can then walk along the E_x surface toward the product molecule to see whether reaction barriers, surface crossings, or near-crossings occur. As illustrated in Chapters 6 and 7, the ability to explore excited surfaces in the above manner is essential if one hopes to predict the outcome of photochemically initiated reactions.

The energy of the photon $h\nu = \epsilon_{v'}^x - \epsilon_v^0$ has been treated as being *precisely* determined by the initial ϵ_v^0 and final $\epsilon_{v'}^x$ energies. However, the energy of the absorbing photons may not be precisely determined, owing to the finite bandwidth of the light source or the lifetime broadening of the excited level $\epsilon_{v'}^x$. In that case, the contributions arising from a finite range of frequencies $\nu_0 \pm \Delta\nu$ must be added up. In attempting to guess the molecular geometries at which the excited surface E_x is entered, ν must be allowed to vary (by $\Delta\nu$) about the mean value ν_0. Experiments involving high-resolution monochromators ($\Delta\nu/c \sim 0.1$ cm^{-1}) and sharp vibrational lines $(1/hc)\Delta\epsilon_{v'}^x \sim 1$ cm^{-1} do not produce significant smoothing of the photon energy (i.e., $\Delta\nu$ is small). However, modern picosecond and nanosecond light sources have bandwidths of 33 cm^{-1} and 0.03 cm^{-1}, respectively, and for very short (~ 1–10 picosecond) light pulses, significant uncertainty in ν can occur, which then requires one to consider a spread in ν values in implementing a picture of the photon-absorption event.

Even if a highly frequency-resolved light source is employed, a reasonably short (10^{-11}–10^{-13} sec) lifetime of the final state $\epsilon_{v'}^x$ can give rise to a spread (3–333 cm^{-1}) in the allowed absorption energies. Thus, when considering excited states that decompose on a relatively fast time scale ($< 10^{-10}$ sec), one must again consider a range of ν values.

Appendix C

Review of Point-Group Symmetry Tools

In this appendix it is assumed that the reader is familiar with molecular point groups, symmetry operations, and character tables. Good introductions to these topics can be found in several references (Cotton, 1963; Eyring, Walter and Kimball, 1944; and Wilson, Decius, and Cross, 1955). In this appendix, we shall only review material that is of direct use in solving the problems in the text.

We begin by summarizing the information content of a representative character table. A fairly complete list of character tables is given at the end of this appendix. At its ground-state equilibrium geometry the ammonia molecule NH_3 belongs to the C_{3v} point group. Its symmetry operations consist of two C_3 rotation axes (rotation by 120° and 240°, respectively, about an axis passing through the nitrogen atom and lying perpendicular to the plane formed by the three hydrogen atoms), three vertical planes σ_v, $\sigma_{v'}$, $\sigma_{v''}$, and the identity operation. These symmetry elements are shown in Figure C-1.

The C_{3v} character table given at the end of the appendix lists the above symmetry operations along with the names of three irreducible representations (A_1, A_2, E) that characterize this point group. Also listed under the title of point group C_{3v} are examples of especially common and important functions [e.g., z, R_z, (x,y)] that transform according to each of the irreducible representations.

To transform according to a certain irreducible representation means that the function, when operated upon by a point-group symmetry operator, yields a linear combination of the functions that transform according to that irreducible representation. For example, a $2p_z$ orbital (z is the C_3 axis of NH_3) on the nitrogen atom belongs to the A_1 representation because it yields 1 times itself when C_3, C_3', σ_v, $\sigma_{v'}$, $\sigma_{v''}$, or the identity operation operates on it. The factor of 1 means that $2p_z$ has A_1 symmetry, since the characters (the numbers listed opposite A_1 and below E, $2C_3$, and $3\sigma_v$ in the C_{3v} character table) of all six symmetry operations are 1 for the A_1 irreducible representation.

The $2p_x$ and $2p_y$ orbitals on the nitrogen atom transform as the E representation, since C_3, C_3', σ_v, $\sigma_{v'}$, $\sigma_{v''}$, and the identity map $2p_x$ and $2p_y$ among one another. For example,

APPENDIX C

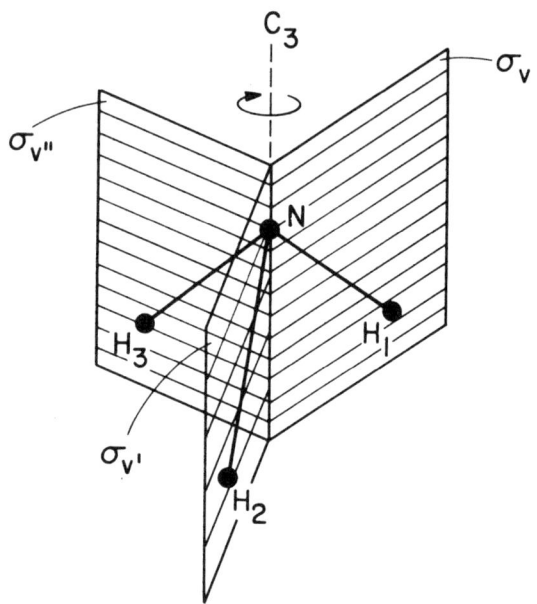

Figure C-1
Symmetry elements of NH$_3$.

$$C_3 \begin{pmatrix} 2p_x \\ 2p_y \end{pmatrix} = \begin{pmatrix} \cos 120° & \sin 120° \\ -\sin 120° & \cos 120° \end{pmatrix} \times \begin{pmatrix} 2p_x \\ 2p_y \end{pmatrix}$$

$$C_3' \begin{pmatrix} 2p_x \\ 2p_y \end{pmatrix} = \begin{pmatrix} \cos 240° & \sin 240° \\ -\sin 240° & \cos 240° \end{pmatrix} \times \begin{pmatrix} 2p_x \\ 2p_y \end{pmatrix}$$

$$E \begin{pmatrix} 2p_x \\ 2p_y \end{pmatrix} = \begin{pmatrix} 1 & 0 \\ 0 & 1 \end{pmatrix} \times \begin{pmatrix} 2p_x \\ 2p_y \end{pmatrix}$$

$$\sigma_v \begin{pmatrix} 2p_x \\ 2p_y \end{pmatrix} = \begin{pmatrix} -1 & 0 \\ 0 & +1 \end{pmatrix} \times \begin{pmatrix} 2p_x \\ 2p_y \end{pmatrix}$$

$$\sigma_{v'} \begin{pmatrix} 2p_x \\ 2p_y \end{pmatrix} = \begin{pmatrix} \frac{1}{2} & -\frac{\sqrt{3}}{2} \\ -\frac{\sqrt{3}}{2} & -\frac{1}{2} \end{pmatrix} \times \begin{pmatrix} 2p_x \\ 2p_y \end{pmatrix}$$

and

$$\sigma_v'' \begin{pmatrix} 2p_x \\ 2p_y \end{pmatrix} = \begin{pmatrix} \frac{1}{2} & \frac{\sqrt{3}}{2} \\ \frac{\sqrt{3}}{2} & -\frac{1}{2} \end{pmatrix} \times \begin{pmatrix} 2p_x \\ 2p_y \end{pmatrix}.$$

The 2×2 matrices, which indicate how each symmetry operation maps $2p_x$ and $2p_y$ into some combinations of $2p_x$ and $2p_y$, are called the representation matrices (RM) for that particular operation and for this particular irreducible representation. For example,

$$\begin{pmatrix} \frac{1}{2} & \frac{\sqrt{3}}{2} \\ \frac{\sqrt{3}}{2} & -\frac{1}{2} \end{pmatrix}$$

is $RM_E(\sigma_v')$. The traces (sums of the diagonal elements) of these matrices are called characters (e.g., $\chi_E(\sigma_v')$) and are the entries in the character tables.

A shortcut device exists for evaluating the trace of the representation matrices (that is, for computing the characters). The diagonal elements of the representation matrices are the projections along each orbital of the effect of the symmetry operation acting on that orbital. For example, a diagonal element of the C_3 matrix is the component of $C_3 2p_y$ along the $2p_y$ direction. More rigorously, it is $\int 2p_y^* C_3 2p_y \, d\mathbf{r}$. Thus, the character of the C_3 matrix is the sum of $\int 2p_y^* C_3 2p_y \, d\mathbf{r}$ and $\int 2p_x^* C_3 2p_x \, d\mathbf{r}$. In general, the character χ of a symmetry operation S can be computed by allowing S to operate on each orbital ϕ_i, projecting $S\phi_i$ along ϕ_i (forming $\int \phi_i^* S \phi_i \, d\mathbf{r}$), and summing these terms, $\Sigma_i \int \phi_i^* S \phi_i \, d\mathbf{r} = \chi(S)$. If these rules are applied to the $2p_x$ and $2p_y$ orbitals of nitrogen within the C_{3v} point group, then

$$\chi(E) = 2, \quad \chi(C_3) = \chi(C_3') = -1, \quad \text{and}$$

$$\chi(\sigma_v) = \chi(\sigma_v') = \chi(\sigma_v'') = 0.$$

This set of characters agrees with those of the E representation for the C_{3v} point group, so $2p_x$ and $2p_y$ belong to or transform as the E representation. This is why (x, y) is to the left of the row of characters for the E representation in the C_{3v} character table.

In similar fashion, the C_{3v} character table states that $d_{x^2-y^2}$ and d_{xy} orbitals on nitrogen transform as E, as do d_{xz} and d_{yz}, but d_{z^2} transforms as A_1.

To illustrate a somewhat more complicated situation, we consider how the three $1s_H$ orbitals on the hydrogen atoms transform. Using the shortcut rule just described, the traces (characters) of the 3×3 representation matrices formed are computed by allowing E, $2C_3$, and $3\sigma_v$ to operate on $1s_{H_1}$, $1s_{H_2}$, and $1s_{H_3}$. The resulting characters are $\chi(E) = 3$, $\chi(C_3) = \chi(C_{3'}) = 0$, and $\chi(\sigma_v) = \chi(\sigma_{v'}) = \chi(\sigma_{v''}) = 1$. The C_{3v} character table shows that these characters $(3,0,1)$ do not match the characters of any *one irreducible* representation, though the sums of the characters of the A_1 and E representations do give these characters. Hence, the hydrogen $1s_H$ orbital set forms a *reducible* representation consisting of the sum of A_1 and E. This means that the three $1s_H$ orbitals can be combined to yield one orbital of A_1 symmetry and a *pair* that each transforms according to the E representation.

To generate the A_1 and E symmetry-adapted orbitals, the symmetry-projection operators P_E and P_{A_1} are used. These operators are given in terms of linear combinations of products of characters times elementary symmetry operations as follows:

$$P_{A_1} = \sum_S \chi_{A_1}(S) S \tag{C.1}$$

$$P_E = \sum_S \chi_E(S) S \tag{C.2}$$

The result of applying P_{A_1} to, say, $1s_{H_1}$ is

$$P_{A_1} 1s_{H_1} = 1s_{H_1} + 1s_{H_2} + 1s_{H_3} + 1s_{H_2} + 1s_{H_3} + 1s_{H_1}$$

$$= 2(1s_{H_2} + 1s_{H_2} + 1s_{H_3}) \equiv \phi_{A_1},$$

which is an (unnormalized) orbital having A_1 symmetry. Clearly, this same ϕ_{A_1} would be generated by P_{A_1} acting on $1s_{H_2}$ or $1s_{H_3}$. Hence, only one A_1 orbital exists.

Likewise,

$$P_E 1s_{H_1} = (2 \times 1s_{H_1}) - 1s_{H_2} - 1s_{H_3} \equiv \phi_{E,1}$$

which is *one* of the symmetry-adapted orbitals having E symmetry. The other E orbital can be obtained by allowing P_E to act on $1s_{H_2}$ or $1s_{H_3}$:

$$P_E 1s_{H_2} = (2 \times 1s_{H_2}) - 1s_{H_1} - 1s_{H_3} \equiv \phi_{E,2}$$

$$P_E 1s_{H_3} = (2 \times 1s_{H_3}) - 1s_{H_1} - 1s_{H_2} \equiv \phi_{E,3}.$$

It might seem as though *three* orbitals having E symmetry were generated, but only two of these are really independent functions. For example, $\phi_{E,3}$ can be expressed in terms of $\phi_{E,1}$ and $\phi_{E,2}$ as

$$\phi_{E,3} = -(\phi_{E,1} + \phi_{,2}).$$

Thus, only $\phi_{E,1}$ and $\phi_{E,2}$ are needed to span the two-dimensional space of the E representation.

In summary, a given set of atomic orbitals $\{\phi_i\}$ can be used as a basis for the symmetry operations of the point group of the molecule. The characters $\chi(S)$ belonging to the operations S of this point group can be found by summing the integrals $\int \phi_i^* S \phi_i \, d\mathbf{r}$ over all the atomic orbitals. The resultant characters will, in general, be reducible to a combination of the characters of the irreducible representations $\chi_i(S)$. To decompose the characters $\chi(S)$ of the reducible representation to a sum of characters $\chi_i(S)$ of the irreducible representation $\chi(S) = \Sigma_i n_i \chi_i(S)$, it is necessary to determine how many times, n_i, the ith irreducible representation occurs in the reducible representation. The expression for n_i is (Cotton, (1963))

$$n_i = \frac{1}{g} \sum_S \chi(S) \chi_i(S) \tag{C.3}$$

in which g is the order of the point group—that is, g is simply the total number of symmetry operations in the group (e.g., $g = 6$ for C_{3v}). The reducible representation $\chi(E) = 3$, $\chi(C_3) = 0$, and $\chi(\sigma_v) = 1$ formed by the three $1s_H$ orbitals discussed above can be decomposed as follows:

$$n_{A_1} = \frac{1}{6}(3 \cdot 1 + 2 \cdot 0 \cdot 1 + 3 \cdot 1 \cdot 1) = 1$$

$$n_{A_1} = \frac{1}{6}(3 \cdot 1 + 2 \cdot 0 \cdot 1 + 3 \cdot 1 \cdot (-1)) = 0$$

$$n_E = \frac{1}{6}(3 \cdot 2 + 2 \cdot 0 \cdot (-1) + 3 \cdot 1(0)) = 1,$$

These equations state that the three $1s_H$ orbitals can be combined to give one A_1 orbital and (since E is degenerate), one *pair* of E orbitals, as established above. With knowledge of the n_i, the symmetry-adapted orbitals can be formed by allowing the projectors

$$P_i = \sum_i \chi_i(S) S \tag{C.4}$$

to operate on each of the primitive atomic orbitals. How this is carried out was illustrated for these $1s_H$ orbitals after equation C.2. These tools allow a symmetry decomposition of any set of atomic orbitals into appropriate symmetry-adapted orbitals.

Before considering other concepts and group-theoretical machinery, it should be pointed out that these same tools can also be used in symmetry analysis of the translational, vibrational, and rotational motions of a molecule. The twelve motions of NH_3 (three translations, three rotations, six vibrations) can be described in terms of combinations of displacements of each of the four atoms in each of three (x, y, z) directions. Hence, unit vectors placed on each atom directed in the x, y, and z directions form a basis for action by the operations (S) of the point group. In the case of NH_3, the characters of the resultant 12×12 representation matrices form a reducible representation in the C_{2v} point group: $\chi(E) = 12$, $\chi(C_3) = \chi(C_{3'}) = 0$, $\chi(\sigma_v) = \chi(\sigma_{v'}) = \chi(\sigma_{v''}) = 2$. This representation can be decomposed as follows:

$$n_{A_1} = \frac{1}{6}[1 \cdot 1 \cdot 12 + 2 \cdot 1 \cdot 0 + 3 \cdot 1 \cdot 2] = 3$$

$$n_{A_2} = \frac{1}{6}[1 \cdot 1 \cdot 12 + 2 \cdot 1 \cdot 0 + 3 \cdot (-1) \cdot 2] = 1$$

$$n_E = \frac{1}{6}[1 \cdot 2 \cdot 12 + 2 \cdot (-1) \cdot 0 + 3 \cdot 0 \cdot 2] = 4.$$

From the information on the left side of the C_{3v} character table, translations of all four atoms in the z, x and y directions transform as $A_1(z)$ and $E(x, y)$, respectively, whereas rotations about the $z(R_z)$, $x(R_x)$, and $y(R_y)$ axes transform as A_2 and E. Hence, of the twelve motions, three translations have A_1 and E symmetry and three rotations have A_2 and E symmetry. This leaves six vibrations, of which two have A_1 symmetry, none have A_2 symmetry, and two (pairs) have E symmetry. We could evaluate the symmetry-adapted vibrations and rotations by allowing symmetry-projection operators of the irreducible-representation symmetries to operate on various elementary cartesian (x, y, z) atomic displacement vectors. Both Cotton (1963) and Wilson, Decius and Cross (1955) show in detail how this is accomplished.

We now return to the symmetry analysis of atomic orbitals by considering how the symmetries of individual orbitals give rise to symmetry characteristics of orbital products. Such knowledge is important because one is routinely faced with constructing symmetry-adapted electronic configurations that consist of products of N individual orbitals. A point-group symmetry operator S, when acting on such a product of orbitals, gives the product of S acting on each of the individual orbitals

$$S\phi_1\phi_2\phi_3 \cdots \phi_N = (S\phi_1)(S\phi_2)(S\phi_3) \cdots (S\phi_N). \tag{C.5}$$

For example, reflection of an N-orbital product through the σ_v plane in NH_3 utilizes reflection operations for all N electrons.

Just as the atomic orbitals formed a basis for action of the point-group operators, the configurations (N-orbital products) form a basis for the action of these same point-group operators. Hence, the various electronic configurations (orbital occupancies) can be treated as functions on which S operates, and the machinery illustrated earlier for decomposing orbital symmetry can then be used to carry out a symmetry analysis of configurations. However, another shortcut makes this task easier. Since the individual orbitals $\{\phi_i, i = 1, \ldots, M\}$ transform according to irreducible representations, we form the direct product of the symmetries of the N orbitals that appear in any configuration. This direct product can then be symmetry-analyzed in a straightforward manner, as discussed earlier. For example, if one is interested in knowing the symmetry of an orbital product involving $a_1^2 a_2^2 e^2$ occupancy in C_{3v} symmetry, the procedure used is the following. For each of the six symmetry operations in the C_{3v} point group, the *product* of the characters associated with each of the *six* spin orbitals (orbital multiplied by α or β spin) is formed

$$\chi(S) = \prod_{i=1}^{6} \chi_i(S) = \chi_{A_1}^2(S)\chi_{A_2}^2(S)\chi_E^2(S). \tag{C.6}$$

In the specific case considered here, $\chi(E) = 4$, $\chi(C_3) = 1$, and $\chi(\sigma_v) = 0$. Notice that the contributions of any doubly occupied nondegenerate orbitals (e.g., a_1^2, and a_2^2) to these direct-product characters $\chi(S)$ are unity because for *all* operators S, $\chi_k^2(S) = 1$ for any nondegenerate irreducible representation k. As a result, only the singly occupied or degenerate orbitals need to be considered when forming the characters of the reducible direct-product representation $\chi(S)$. In this example, the direct-product characters can be determined from the characters $\chi_E(S)$ of the two active (non-closed-shell) orbitals—the e^2 orbitals. That is, $\chi(S) = \chi_E(S) \cdot \chi_E(S)$.

From the direct-product characters $\chi(S)$ that belong to a particular electronic configuration (e.g., $a_1^2 a_2^2 e^2$), one must still decompose this list of characters into a sum of irreducible characters using equation C.3. For the example at hand, the direct-product characters $\chi(S)$ decompose into one A_1, one A_2, and one E representation. This means that the e^2 configuration contains A_1, A_2, and E symmetry elements. The e^2 configuration contains all determinants that can be formed by placing two electrons into the *pair* of degenerate orbitals. There are six such determinants. In Chapter 4 we show how to form combinations of these Slater-determinant wavefunctions that display these pure (A_1, A_2, and E) symmetries and that possess either singlet or triplet spin (which are the only possibilities for the two e^2 electrons).

In summary, we have reviewed how to make a symmetry decomposition of a basis of atomic orbitals into their irreducible representation components. This tool is most helpful when constructing the orbital-correlation diagrams that form the basis of the Woodward-Hoffman rules. We also learned how to form the direct-product symmetries that arise when considering configurations that consist of products of symmetry-adapted spin orbitals. This step is essential for the construction of configuration- and state-correlation diagrams upon which one ultimately bases a prediction about whether a reaction is allowed or forbidden.

C_1	E
A	1

C_2			E	C_2
x^2, y^2, z^2, xy	R_s, z	A	1	1
xz, yz	x, y R_x, R_y	B	1	-1

C_3			E	C_3	C_3^2	
$x^2 + y^2, z^2$	R_s, z	A	1	1	1	
(xz, yz)	(x, y)	E	1	ω	ω^2	$(\omega = e^{2\pi i/3})$
$(x^2 - y^2, xy)$	(R_x, R_y)		1	ω^2	ω	

C_4			E	C_2	C_4	C_4^3
$x^2 + y^2, z^2$	R_s, z	A	1	1	1	1
$x^2 - y^2, xy$		B	1	1	-1	-1
(xz, yz)	(x, y) (R_x, R_y)	E	1	-1	i	$-i$
			1	-1	$-i$	i

	C_5		E	C_5	C_5^2	C_5^3	C_5^4	
x^2+y^2, z^2	R_s, z	A	1	1	1	1	1	
(xz, yz)	(x, y)	E'	1	ω	ω^2	ω^3	ω^4	
	(R_x, R_y)		1	ω^4	ω^3	ω^2	ω	$(\omega = e^{2\pi i/5})$
(x^2-y^2, xy)		E''	1	ω^2	ω^4	ω	ω^3	
			1	ω^3	ω	ω^4	ω^2	

	C_6		E	C_6	C_3	C	C_3^2	C_6^5	
x^2+y^2, z^2	R_z, z	A	1	1	1	1	1	1	
		B	1	-1	1	-1	1	-1	
(xz, yz)	(x, y)	E'	1	ω	ω^2	ω^3	ω^4	ω^5	$(\omega = e^{2\pi i/6})$
	(R_x, R_y)		1	ω^5	ω^4	ω^3	ω^2	ω	
(x^2-y^2, xy)		E''	1	ω^2	ω^4	1	ω^2	ω^4	
			1	ω^4	ω^2	1	ω^4	ω^2	

	C_{2v}		E	C_2	σ_v	σ_v'
x^2, y^2, z^2	z	A_1	1	1	1	1
xy	R_s	A_2	1	1	-1	-1
xz	R_y, x	B_1	1	-1	1	-1
yz	R_x, y	B_2	1	-1	-1	1

	C_{3v}		E	$2C_3$	$3\sigma_v$
x^2+y^2, z^2	z	A_1	1	1	1
	R_s	A_2	1	1	-1
(x^2-y^2, xy)	(x, y)	E	2	-1	0
(xz, yz)	(R_x, R_y)				

	C_{4v}		E	C_2	$2C_4$	$2\sigma_v$	$2\sigma_d$
x^2+y^2, z^2	z	A_1	1	1	1	1	1
	R_s	A_2	1	1	1	-1	-1
x^2-y^2		B_1	1	1	-1	1	-1
xy		B_2	1	1	-1	-1	1
(xz, yz)	(x, y) (R_x, R_y)	E	2	-2	0	0	0

C_{5v}			E	$2C_5$	$2C_5^2$	$5\sigma_v$	
x^2+y^2, z^2	z	A_1	1	1	1	1	
	R_s	A_2	1	1	1	-1	$x = \dfrac{2\pi}{5}$
(xz, yz)	(x, y)	E_1	2	$2\cos x$	$2\cos 2x$	0	
	(R_x, R_y)	E_2	2	$2\cos 2x$	$2\cos 4x$	0	
(x^2-y^2, xy)							

C_{6v}			E	C_2	$2C_3$	$2C_6$	$3\sigma_d$	$3\sigma_v$
x^2+y^2, z^2	z	A_1	1	1	1	1	1	1
	R_z	A_2	1	1	1	1	-1	-1
		B_1	1	-1	1	-1	-1	1
		B_2	1	-1	1	-1	1	-1
(xz, yz)	(x, y)	E_1	2	-2	-1	1	0	0
	(R_x, R_y)							
(x^2-y^2, xy)		E_2	2	2	-1	-1	0	0

C_{1h}			E	σ_h
x^2, y^2, z^2, xy	R_z, x, y	A'	1	1
xz, yz	R_x, R_y, z	A''	1	-1

C_{2h}			E	C_2	σ_h	i
x^2, y^2, z^2, xy	R_z	A_g	1	1	1	1
	z	A_u	1	1	-1	-1
xz, yz	R_x, R_y	B_g	1	-1	-1	1
	x, y	B_u	1	-1	1	-1

$C_{3h} = C_3 \times \sigma_h$			E	C_3	C_3^2	σ_h	S_3	$(\sigma_h C_3^2)$	
x^2+y_2, z^2	R_z	A'	1	1	1	1	1	1	
	z	A''	1	1	1	-1	-1	-1	
(x^2-y^2, xy)	(x, y)	E'	1	ω	ω^2	1	ω	ω^2	$(\omega = e^{2\pi i/3})$
			1	ω^2	ω	1	ω^2	ω	
(xz, yz)	(R_x, R_y)	E''	1	ω	ω^2	-1	$-\omega$	$-\omega^2$	
			1	ω^2	ω	-1	$-\omega^2$	$-\omega$	

$$C_{4h} = C_4 \times i$$

$$C_{5h} = C_5 \times \sigma_h$$

$$C_{6h} = C_6 \times i$$

		S_2		E	i
x^2, y^2, z^2, xy	R_x, R_y, R_z		A_g	1	1
xz, yz					
	x, y, z		A_u	1	−1

		S_4		E	C_2	S_4	S_4^3
$x^2 + y^2, z^2$	R_z		A	1	1	1	1
	z		B	1	1	−1	−1
(xz, yz)	(x, y)		E	1	−1	i	$-i$
$(x^2 - y^2, xy)$	(R_x, R_y)			1	−1	$-i$	i

$$S_6 = C_3 \times i$$

		D_2		E	C_2^z	C_2^y	C_2^x
x^2, y^2, z^2			A_1	1	1	1	1
xy	R_z, z		B_1	1	1	−1	−1
xz	R_y, y		B_2	1	−1	1	−1
yz	R_x, x		B_3	1	−1	−1	1

		D_3		E	$2C_3$	$3C_2'$
$x^2 + y^2, z^2$			A_1	1	1	1
	R_z, z		A_2	1	1	−1
(xz, yz)	(x, y)		E	2	−1	0
$(x^2 - y^2, xy)$	(R_x, R_y)					

APPENDIX C

	D_4		E	C_2	$2C_4$	$2C_2'$	$2C_2''$
x^2+y^2, z^2		A_1	1	1	1	1	1
	R_z, z	A_2	1	1	1	-1	-1
		B_1	1	1	-1	1	-1
		B_2	1	1	-1	-1	1
(xz, yz) (x^2-y^2, xy)	(x, y) (R_x, R_y)	E	2	-2	0	0	0

	D_5		E	$2C_5$	$2C_5^2$	$5C_2'$	
x^2+y^2, z^2		A_1	1	1	1	1	
	R_z, z	A_2	1	1	1	-1	
(xz, yz)	(x, y) (R_x, R_y)	E_1	2	$2\cos x$	$2\cos 2x$	0	$x = \dfrac{2\pi}{5}$
(x^2-y^2, xy)		E_2	2	$2\cos 2x$	$2\cos 4x$	0	

	D_6		E	C_2	$2C_3$	$2C_6$	$3C_2'$	$3C_2''$
x^2+y^2, z^2		A_1	1	1	1	1	1	1
	R_z, z	A_2	1	1	1	1	-1	-1
		B_1	1	-1	1	-1	1	-1
		B_2	1	-1	1	-1	-1	1
(xz, yz)	(x, y) (R_x, R_y)	E_1	2	-2	-1	1	0	0
(x^2-y^2, xy)		E_2	2	2	-1	-1	0	0

	D_{2d}		E	C_2	$2S_4$	$2C_2'$	$2\sigma_d$
x^2+y^2, z^2		A_1	1	1	1	1	1
	R_z	A_2	1	1	1	-1	-1
x^2-y^2		B_1	1	1	-1	1	-1
xy		B_2	1	1	-1	-1	1
	z						
(xz, yz)	(x, y) (R_x, R_y)	E	2	-2	0	0	0

$$D_{3d} = D_3 \times i$$

$$D_{2h} = D_2 \times i$$

$D_{3h} = D_3 \times \sigma_h$			E	σ_h	$2C_3$	$2S_3$	$3C_2'$	$3\sigma_v$
x^2+y^2, z^2		A_1'	1	1	1	1	1	1
	R_z	A_2'	1	1	1	1	-1	-1
		A_1''	1	-1	1	-1	1	-1
	z	A_2''	1	-1	1	-1	-1	1
(x^2-y^2, xy)	(x, y)	E'	2	2	-1	-1	0	0
(xz, yz)	(R_x, R_y)	E''	2	-2	-1	1	0	0

$$D_{4h} = D_4 \times i$$

$$D_{5h} = D_5 \times \sigma_h$$

$$D_{6h} = D_6 \times i$$

	T		E	$3C_2$	$4C_3$	$4C_3'$	
Active		A	1	1	1	1	
Active		E	1	1	ω	ω^2	
			1	1	ω^2	ω	$\omega = e^{2\pi i/3}$
Active	(R_x, R_y, R_z) (x, y, z)	T	3	-1	0	0	

$$T_h = T \times i$$

	O		E	$8C_3$	$3C_2$	$6C_2$	$6C_4$
Active		A_1	1	1	1	1	1
Inactive		A_2	1	1	1	-1	-1
Active		E	2	-1	2	0	0
Active	(R_x, R_y, R_z) (x, y, z)	T_1	3	0	-1	-1	$+1$
Active		T_2	3	0	-1	$+1$	-1

$$O_h = O \times i$$

	T_d		E	$8C_3$	$3C_2$	$6\sigma_d$	$6S_4$
Active		A_1	1	1	1	1	1
Inactive		A_2	1	1	1	-1	-1
Active		E	2	-1	2	0	0
Active	(R_x, R_y, R_z)	T_1	3	0	-1	1	-1
Active	(x, y, z)	T_2	3	0	-1	-1	1

		$C_{\infty v}$	E	$2C_\varphi$	σ_v
x^2+y^2, z^2	z	A_1	1	1	1
	R_z	A_2	1	1	-1
(xz, yz)	(x, y) (R_x, R_y)	E_1	2	$2\cos\varphi$	0
(x^2-y^2, xy)		E_2	2	$2\cos 2\varphi$	0
		$\cdots$			

		$D_{\infty h}$	E	$2C_\varphi$	C_2'	i	$2iC_\varphi$	iC_2'
x^2+y^2, z^2		A_{1g}	1	1	1	1	1	1
		A_{1u}	1	1	1	-1	-1	-1
		A_{2g}	1	1	-1	1	1	-1
	z	A_{2u}	1	1	-1	-1	-1	1
(xz, yz)	(R_x, R_y)	E_{1g}	2	$2\cos\varphi$	0	2	$2\cos\varphi$	1
	(x, y)	E_{1u}	2	$2\cos\varphi$	0	-2	$-2\cos\varphi$	0
(x^2-y^2, xy)		E_{2g}	2	$2\cos 2\varphi$	0	2	$2\cos 2\varphi$	0
		E_{2u}	2	$2\cos 2\varphi$	0	-2	$-2\cos 2\varphi$	0
		$\cdots$						

Answers

Problems Relating to Thermal Processes (Chapter 4)

1. The relevant HOMO and LUMO of the ten-membered ring are the antisymmetric σ_A bonding and symmetric antibonding σ_S^* orbitals involving the CH bonds. The HOMO and LUMO of the smaller ring are the π_S and π_A^* orbitals, which are symmetric and antisymmetric, respectively, under reflection through the plane of symmetry that is preserved throughout the reaction. Suprafacial attack would result in favorable HOMO-LUMO interactions ($\sigma_A \leftrightarrow \pi_A^*$ and $\sigma_S^* \leftrightarrow \pi_S$), whereas antarafacial attack would not.

The bond-symmetry rule also indicates that suprafacial attack is allowed because the occupied orbitals of the reactants (σ_S, σ_A, and π_S) match in symmetry those of the products (π_S, σ_S, and σ_A), where the π bond is now in the ten-membered ring, and σ_S and σ_A now refer to CH bonds in the smaller ring.

In applying the Dewar-Zimmerman method, one finds for suprafacial attack, a Hückel transition state having six electrons (two CH bonds and one π bond). Again, the suprafacial attack is predicted to be allowed.

2. Using the bond-symmetry rule, one sees that the occupied active orbitals of the cyclopenteneone are the symmetric σ_S and antisymmetric σ_A CC bonding orbitals and the symmetric p_S orbital. The plane of symmetry used to make these labels is the only one that persists throughout the reaction path. In the products the active orbitals are the CO lone pair on carbon, which is symmetric $\sigma(CO)_S$, and the two occupied π orbitals of 1,3-butadiene, which are symmetric π_S and antisymmetric π_A, respectively. The reactant and product orbitals match in symmetry; hence, the decomposition reaction should be thermally allowed.

In the other case, the relevant occupied orbitals of the cyclohexadieneone are the two CC bonds σ_S and σ_A and the two π bonds π_S and π_S' (both of which are symmetric). In the products, the orbitals are $\sigma_S(CO)$ and the three occupied orbitals of benzene, which are π_S, π_S', and π_A (see section 7.6). Again, the bond-symmetry rule indicates that the thermal decomposition reaction is allowed.

3. This reaction is symmetry-forbidden—it is nothing but two independent $[2_s + 2_s]$ cycloaddition reactions. Such reactions were shown to be forbidden in sections 4.4 and 4.8. The reason this problem might lead to confusion is that when using the orbital- and configuration-correlation diagram method, one is tempted to connect the $\pi_{12} - \pi_{56}$ symmetry-adapted orbital, which is antisymmetric under the plane $M1$, with the $\sigma_{38} - \sigma_{47}$ orbital, which is also odd under $M1$. However, as becomes clear when one utilizes the orbital-following device, $\pi_{12} - \pi_{56}$ and $\sigma_{38} - \sigma_{47}$ cannot be so connected—they belong to totally distinct portions on this molecule.

By placing two forbidden $[2_s + 2_s]$ reactions in close spatial proximity, one might incorrectly correlate the reactant and product orbitals. It is not proper to correlate orbitals that are localized on one part of the reactant molecule with those that belong to a different part of the product molecule.

4. If n is odd, the H atom can undergo a suprafacial shift to give $HDFC=C(-C=C)_n-CR_1R_2$; when n is even, the antarafacial hydrogen shift to the terminal carbon is allowed. In either case, one obtains two isomers that differ geometrically at the $-CR_1R_2$ end and that are enantiomers at the $HDFC-$ end. The two isomers arise in each case because of the free rotation about the $C-CHR_1R_2$ bond in the reactant molecule.

5. Methyl-group migration is allowed because the orbital of the CH_3 group, which plays a role analogous to that of the $1s$ orbital of the hydrogen atom, has sp^3 character. This orbital has both a positive and a negative lobe. By connecting its positive lobe to the orbital of the neighboring carbon atom and its negative lobe to the p_π orbital of the terminal carbon atom, one achieves a Möbius transition state having four electrons. Thus, the suprafacial methyl-group shift is allowed. Of course, the configuration of the substituents around the methyl group is inverted once the transfer to the 3-carbon takes place.

6. Denoting the three orbitals of the hydrogen atoms by $1s_{HA}$, $1s_{HB}$, and $1s_{HC}$ and applying the a_1 and e symmetry projectors (see Appendix C), one obtains the following (unnormalized) symmetry-adapted orbitals:

$$X_{a_1} = (1s_{HA} + 1s_{HB} + 1s_{HC})$$

$$X_e = \begin{cases} 2 \cdot 1s_{HA} - 1s_{HB} - 1s_{HC} \\ 2 \cdot 1s_{HB} - 1s_{HA} - 1s_{HC} \end{cases}$$

The four nitrogen orbitals can also be symmetry-projected:

$$X'_{a_1} = 2s_N$$

$$X''_{a_1} = 2p_{zN}$$

$$X'_e = \begin{cases} 2p_{xN} \\ 2p_{yN} \end{cases}$$

(The z axis is chosen to be the 3-fold symmetry axis of the molecule.)

The three a_1 atomic orbitals combine to yield bonding (ϕ_1), nonbonding (ϕ_4) and antibonding (ϕ_7) molecular orbitals having a_1 symmetry. Likewise, the two pairs of e orbitals combine to give pairs of bonding (ϕ_2,ϕ_3) and antibonding (ϕ_5,ϕ_6) molecular orbitals having e symmetry.

The ground state of NH$_3$ has an electronic wavefunction that is dominated by the configuration $\phi_1^2\phi_2^2\phi_3^2\phi_4^2$ (the $1s_N^2$ electrons are neglected). This configuration has $^1A_1$ symmetry. The singly excited configuration $\phi_1^2\phi_2^2\phi_3\phi_4^2\phi_5$ gives rise to singlet and triplet states corresponding to all symmetries contained in the direct product $e \times e = e + a_1 + a_2$ (see Appendix C). Of these, the E state would be first-order Jahn-Teller unstable, whereas the other two (A_1 and A_2) are not. The A_1 and A_2 states could be second-order (actually pseudo-) Jahn-Teller unstable through coupling, via a distortion of e symmetry, with the E state. The other singly excited configuration $\phi_1^2\phi_2^2\phi_3\phi_4^2\phi_7$ has $e \times a_1 = e$ symmetry. This E state should be first-order Jahn-Teller unstable with respect to distortions of $E \times E = E + A_1 + A_2$ symmetry. Of these, the only vibrations of NH$_3$ have A_1 and E symmetry. The A_1 vibrations would not remove the degeneracy because they preserve the symmetry of the molecule; hence, only the distortion of E symmetry will be effective.

Problems on Photochemistry (Chapters 5–7)

The answers to problem 1 are given in the excellent book *Problems in Quantum Chemistry* by P. Jørgensen and J. Oddershede (Addison-Wesley, Reading, Mass., 1983). On page 238 of this book a discussion of the problem is given, as well as references to the experimental literature relating to this very interesting case.

Complete answers to the questions posed in problem 2 are probably not attainable at this time. Much debate remains about what is really happening in the photochemistry of formaldehyde. For this reason, it is best to attempt to relate your answers to this problem to some of the best treatments of formaldehyde photochemistry, which are contained in the following references: J. C. Weisshaar and C. B. Moore, (1980), *J. Chem. Phys.,* 72, 5415; H. L. Selzle and E. W. Schlag, (1979), *Chem. Phys.,* 43, 111; D. F. Heller, M. L. Elert, and W. M. Gelbart, (1978), *J. Chem. Phys.,* 69, 4061; J. D. Goddard and H. F. Schaefer III, (1979), *J. Chem. Phys.,* 70, 5117; and many other references contained in these papers.

References

Chapter 1

Bordon, W. T. (1975). *Molecular Orbital Theory for Organic Chemists*. Prentice-Hall.
Eyring, H., J. Walter, and G. E. Kimball. (1944). *Quantum Chemistry*. Wiley.
Fleming, I. (1976). *Frontier Orbitals and Organic Chemical Reactions*. John Wiley and Sons.
Mead, C. A. (1979). *J. Chem. Phys.*, 70, 2276.
Pack R. T. and J. O. Hirschfelder. (1968). *J. Chem. Phys.*, 49, 4009; *ibid.*, (1970), 52, 528.
Pearson, R. G. (1976). *Symmetry Rules for Chemical Reactions*. Wiley.
Wilson, Jr., E. B., J. C. Decius, and P. C. Cross. (1955). *Molecular Vibrations*. McGraw-Hill.
Wolniewicz, L. and W. Kolos. (1963). *Rev. Mod. Phys.*, 35, 473; *J. Chem. Phys.*, 41, 3663, 3674 (1964); 43, 2429 (1965).
Woodward, R. B. and R. Hoffmann. (1970). *The Conservation of Orbital Symmetry*. Verlag Chemie Gmbh., Weinbeim/Bergstrasse.

Chapter 2

Arfken, G. (1970). *Mathematical Methods for Physics*. Academic Press.
Cerjan, C. and W. H. Miller. (1981). *J. Chem. Phys.*, 75, 2800.
Cotton, F. A. (1963). *Chemical Applications of Group Theory*. Interscience.
Herzberg, G. (1966). *Electronic Spectra of Polyatomic Molecules*, Van Nostrand.
Jørgensen, P. and J. Simons. (1981). *Second Quantization Based Methods in Quantum Chemistry*. Academic Press.
Simons, J., P. Jørgensen, H. Taylor, and J. Ozment. (1983). *J. Phys. Chem.*, 87, 2745 (1983).
Wigner, E. P. (1959). *Group Theory*. Academic Press.
Wilson Jr., E. B., J. C. Decius, and P. C. Cross. (1955). *Molecular Vibrations*. McGraw-Hill.

Chapter 3

Condon E. V. and G. H. Shortley. (1957). *The Theory of Atomic Spectra*. Cambridge University Press.
Cook, D. B. (1978). *Structures and Approximations to Electrons in Molecules*. Ellis Horwood, Sussex, England.

Cotton, F. A. (1963). *Chemical Applications of Group Theory*. Interscience.
Dunning, Jr., T. H. (1970). *J. Chem. Phys.*, 53, 2823; 55, 716 (1971).
Dunning, Jr., T. H. and P. J. Hay. (1977). Ch. I in *Modern Theoretical Chemistry*. Ed. by H. F. Schaefer III. Plenum.
Pilar, F. L. (1968). *Elementary Quantum Chemistry*. McGraw-Hill.
Roothaan, C. C. J. (1951). *Rev. Mod. Phys.*, 23, 69.
Shavitt, I. (1977). In *Modern Theoretical Chemistry*. Ed. by H. F. Schaefer III. Plenum.

Chapter 4

Benson, S. W. (1960). *The Foundations of Chemical Kinetics*. McGraw-Hill.
Cotton, F. A. (1963). *Chemical Applications of Group Theory*. Interscience.
Dewar, M. J. (1966). *Tetrahedron*, Suppl. 8, 75.
Fukui, H. (1971). *Acct. Chem. Res.*, 4, 57.
Pearson, R. G. (1976). *Symmetry Rules for Chemical Reactions*. Wiley.
Pilar, F. L. (1968). *Elementary Quantum Chemistry*. McGraw-Hill.
Woodward, R. B. and R. Hoffman. (1970). *The Conservation of Orbital Symmetry*, Verlag Chemie, Gmbh. Weinbeim/Bergstrasse.
Zimmerman, H. E. (1966). *J. Amer. Chem. Soc.*, 88, 1564, 1566.

Chapter 5

Beer, M. and H. C. Longuet-Higgins. (1955). *J. Chem. Phys.*, 23, 1390.
Kasha, M. (1950). *Disc. Faraday Soc.*, 9, 14.
McMurchie, L. E. and E. R. Davidson. (1977). *J. Chem. Phys.*, 66, 2959.
Michl, J. (1972). *Molec. Photochem.* 4, 243; (1974) *Topics in Current Chemistry*, 46, 1; (1975) *Pure Applied Chem.*, 41, 507.
Pearson, R. G. (1976). *Symmetry Rules for Chemical Reactions*. Wiley.
Rice, S. A. (1971). *Adv. Chem. Phys.*, 21, 153.
Turro, N. J. (1978). *Modern Molecular Photochemistry*. Benjamin/Cummings.

Chapter 6

Berry, R. S. (1966). *J. Chem. Phys.*, 45, 1278.
Eyring, H., J. Walter, and G. E. Kimball. (1944). *Quantum Chemistry*. Wiley.
Lin, S. H. (1980). *Radiationless Transitions*. Academic Press.
Yardley, J. T. (1980). *Introduction to Molecular Energy Transfer*. Academic Press.

Chapter 7

Cotton, F. A. (1963). *Chemical Applications of Group Theory*. Interscience.
Pearson, R. G. (1976). *Symmetry Rules for Chemical Reactions*. Wiley.

Appendix A

Bartlett, R. J. and D. M. Silver. (1975). *Inter. J. Quantum Chem.*, S9, 183.
Hehre, W. J., R. F. Stewart and J. A. Pople. (1969). *J. Chem. Phys.*, 51, 2657.

Hylleraas, E. A. and B. Undheim. (1930). *Z. Phys.*, 65, 759.
Jørgensen, P. and J. Simons. (1981). *Second Quantization Based Methods in Quantum Chemistry*. Academic Press.
Pauncz, R. (1979). *Spin Eigenfunctions*. Plenum.
Pople, J. A., J. S. Binkley, and R. Seeger, *Inter. J. Quantum Chem.*, S10, 1.
Roothaan, C. C. J. (1960). *Rev. Mod. Phys.*, 32, 179.
Schaefer, H. F. (1972). *The Electronic Structure of Atoms and Molecules*. Addison-Wesley.
Shavitt, I. (1977). *Modern Theoretical Chemistry*, Vol. 3. Ed. by H. F. Schaefer. Plenum.

Appendix B

Simons, J. (1982). *J. Phys. Chem.*, 86, 3615.

Appendix C

Cotton, F. A. (1963). *Chemical Applications of Group Theory*. Interscience.
Eyring, H., J. Walter, and G. E. Kimball. (1944). *Quantum Chemistry*. Wiley.
Wilson Jr., E. B., J. C. Decius, and P. C. Cross. (1955). *Molecular Vibrations*. McGraw-Hill.

Index

activated complex. *See* Transition state
adiabatic approximation, 6
angular-momentum operators, 114–117
antarafacial, 59
atomic-orbital basis
    Gaussian-type, 25, 125–127
        contracted, 126
    integrals, 25
    Slater-type, 25, 125–127
avoided configuration crossings, 38, 43
    strongly, 52
    weakly, 38, 43, 46

bond-symmetry rule, 46–50
bonding-antibonding orbital
    mixing, 52–54, 102–104
Born-Oppenheimer approximation, 6, 83, 123, 138
    corrections to, 7, 84
Brillouin theorem, 129

character table, 141–153
    list of, 148–154
characters, 143
clamped-nuclei concept, 4, 123
configuration-correlation diagram (CCD) 29, 37, 43, 48, 97, 100, 101, 106, 109–111, 118
configuration interaction (mixing), 23, 28–30, 128–135
    strength, 128
concerted reaction, 52
conrotatory motion, 50, 54–58, 99–100
cusp condition, 126
cyclic transition states, 65–68, 101–102
cycloaddition reactions, 58

density of states, 85
Diels-Alder reaction, 60, 102–104
direct product, 146–148
    antisymmetric, 45
    symmetric, 45–46
disrotatory motion, 50, 54–58, 99–100

eigenfunctions
    electronic, 5
    total, 5
    vibration-rotation, 6
electric-dipole
    transition moment, 138–139
electrocyclic reaction, 53
energy-digesting modes, 87–91, 93
energy gap, 90

Fermi golden rule, 84
Fock operator, 24, 26, 124
Franck-Condon
    factors, 88, 138
    principle, 79, 98
funnels on potential energy surfaces, 78, 98, 100

Hamiltonian
    electronic, 5, 13–15, 123
    nuclear-motion, 4
    total, 3
Hartree-Fock
    energy, 27–28
    procedure, 24–28, 123–127
        unrestricted, 124
        restricted, 125
        projected, 124

wavefunction, 27-28, 123
highest occupied molecular orbital (HOMO), 54
hopping between potential energy surfaces, 77-78, 111-112, 119-120

intensity borrowing, 139
internal conversion, 77, 83-92, 98
intersystem crossing, 79-80, 83-85, 92-94, 99

Jahn-Teller effect
 first-order, 17, 44
 pseudo, 48, 119-120
 second-order, 17-21
 vibrational distortions in, 44, 46-48

Kasha's rule, 77
Koopmans' theorem, 36, 127

Landau-Zener theory, 91-92, 94
linear combination of atomic orbitals (LCAO) 26, 125
lowest unoccupied molecular orbital (LUMO), 54

many-body perturbation theory, 134-135
molecular orbital, 23
 configurations, 23
 energies, 27, 124
 Hartree-Fock, 24, 123
 occupied, 27
 spin, 26
 symmetry, 24-25
 virtual, 27

occupied-orbital following, 54
one- and two-electron integral transformation, 132
orbital-correlation diagram (OCD), 29, 36, 42, 47, 96, 100, 105, 108, 114
orbital nodal patterns, 55, 64
orbital relaxation, 129

partly classical view of photon absorption, 139-140
photon absorption probability, 138-140
point-group symmetry. See Appendix C
polarized orbital pairs, 129-130
potential energy surface, 3, 5, 11ff
 slopes, 11-17
 curvatures, 11-21
 intersection, 8, 29, 81
 transitions between, 77
perturbation theory, 15-20, 92-94

radiationless transition, 7, 83-94
 rate, 84-94
reaction coordinate, 12
reaction path, 12, 13
reducible representation, 45, 144, 147
 character of, 45, 144, 147
representation matrices, 143

sigmatropic migrations, 63-65
singlet and triplet wavefunctions, 76, 92-94, 114-118
size consistency, 133-135
Slater-Condon rules, 29, 130-132
Slater determinant, 123, 124
spin-orbit operator, 92
spin orbital, 123
spin symmetry
 conservation, 75
 notation (S_0, T_1, S_1), 75
state-correlation diagram, 29, 37, 43
suprafacial, 59
symmetry
 approximate, 29, 39-40, 43, 49
 direct products in, 146-148
 effects on surface curvatures, 17-20
 effects on surface slopes, 16-17
 elements, 141
 of molecular deformations, 15, 16
 of vibrations, 146
 point-group. See Appendix C
 projection operators, 44, 45, 144-145

thermal reaction, 38
transition state, 11, 46